Solid-State Ignition Systems

by

Rudolf F. Graf

George J. Whalen

HOWARD W. SAMS & CO., INC.
THE BOBBS-MERRILL CO., INC.
INDIANAPOLIS · KANSAS CITY · NEW YORK

FIRST EDITION

FIRST PRINTING—1974

Copyright © 1974 by Howard W. Sams & Co., Inc., Indianapolis, Indiana 46268. Printed in the United States of America.

All rights reserved. Reproduction or use, without express permission, of editorial or pictorial content, in any manner, is prohibited. No patent liability is assumed with respect to the use of the information contained herein.

International Standard Book Number: 0-672-21049-5
Library of Congress Catalog Card Number: 74-75277

Preface

The idea of a solid-state ignition system provided good copy for popular magazines in the 1950's—an era of gas-guzzling 400-plus-cubic-inch speed monsters. Super performance and high speed via transistor were the tantalizing qualities then dangled before an affluent postwar car-buying public, and the term "transistor ignition" was magic!

But the magic did not materialize in the fifties. Early transistor ignition systems were tested—and flopped—in the hostile underhood crucible of the automobile! Discarded by car makers and many aftermarket sources, the discredited electronic ignition system limped into a "dark age" that persisted until the mid-1960's. Then, the fortunate advent of rugged silicon semiconductors and the impetus of progressively more demanding governmental emission-control standards fostered a great leap forward in solid-state ignition design.

Suddenly, car makers and electronic-systems manufacturers were back at work on electronic ignition—this time with a more mature purpose than feeding the speed lust of the 1950's. Preservation of markets and clean air mandated reliable ignition. Paradoxically, solid-state ignition could now be made more reliable than the breaker-point ignition systems it had lost out to earlier. Hard on the heels of the environmental push, a gasoline shortage added painful spurs to the need for fuel economy, further solidifying the position of the solid-state ignition system in the modern automobile.

Solid-state ignition is a fact of life in the car of today. There are many types of systems, reflecting the various schools of thought in engineering about what constitutes the "ideal" solid-state ignition system. You will find all the known, commercially available methods of solid-state ignition explored in this book. We have conducted a thorough examination of original equipment designs as well as add-on systems, and we present our findings for your study.

You need not be an engineer to read and enjoy this book. The first chapter describes the phenomenon of spark ignition in easy-to-understand terms. Chapter 2 then relates how the venerable conventional ignition system produces a spark. Chapter 3 discusses semiconductors and how they work in ignition systems. Detailed discussions of solid-state systems are provided in Chapters 4 and 5. For those interested in servicing solid-state ignition systems, Chapters 6 presents comprehensive approaches to servicing and troubleshooting the various system types available today.

We are indebted to many industry sources who so generously contributed time and expertise in order that our work might be complete and authoritative. Most particularly, we wish to extend our sincere thanks to Robert G. Van Houten, Charles Cimilluca, Robert Actis, Jim Stickford, William Stempien, and Charles Mulcahy. Special thanks go to Mrs. John J. Dillon for her meritorious efforts at the typewriter.

RUDOLF F. GRAF
GEORGE J. WHALEN

Contents

CHAPTER 1

IGNITION FUNDAMENTALS 7

    Ignition—The Spark Plug and Energy Conversion

CHAPTER 2

CONVENTIONAL IGNITION 13

    Kettering Ignition—Problems of Conventional Kettering Ignition

CHAPTER 3

A PRACTICAL REVIEW OF SEMICONDUCTOR THEORY 21

    The Semiconductor Junction—The Semiconductor Diode—The Transistor—The Silicon Controlled Rectifier—The Programmable Unijunction Transistor

CHAPTER 4

UNDERSTANDING SOLID-STATE IGNITION SYSTEMS 31

    Inductive-Discharge Systems—Operating Principles of Solid-State Inductive-Discharge Systems—Single-Transistor Contact-Triggered Ignition Circuits—Dual-Transistor Contact-Triggered Ignition Circuits—Contact-Isolator Switching System—Dwell-Extender Systems—Magnetically Triggered Inductive-Discharge Ignition Systems

CHAPTER 5

CAPACITIVE-DISCHARGE IGNITION SYSTEMS 69

How Capacitive-Discharge Ignition Systems Operate—Inverter-Type CD Ignition Systems—Magneto-Type CD Ignition Systems

CHAPTER 6

TROUBLESHOOTING SOLID-STATE IGNITION SYSTEMS 103

Problems Common to Most Systems—Troubleshooting the Ignition Secondary Circuit—Troubleshooting the Ignition Primary Circuit—Contact-Triggered Inductive-Discharge Systems—Magnetically Triggered Inductive-Discharge Systems—Capacitive-Discharge Systems—Magneto-Type System

INDEX 133

CHAPTER 1

Ignition Fundamentals

IGNITION

Ignition has played a key role in motive power ever since Nikolaus August Otto and his partner Eugene Langen devised the first four-stroke-cycle internal combustion engine in the late 1870s (Fig. 1-1). Traditionally, an *electric spark* has been the means used to ignite a

Fig. 1-1. The first successful four-stroke-cycle engine, designed by Nicolaus August Otto.

7

fuel/air charge to start combustion—the liberation of heat energy—which is the actual power source in gas engine operation.

In an Otto-cycle engine, the ignition spark marks the commencement of a rapid oxidant-reductant reaction that liberates heat within each cylinder. Compressed gasoline and oxygen atoms combine so quickly and violently that enormous heat is produced. The heat causes super expansion of combustion gases, which impart a mechanical push against all the surfaces of the combustion chamber. One of these surfaces (the piston or the rotor of a Wankel engine) is free to move and thus "soaks up" the push energy, while also enlarging the space within which the hot gas has been confined. The hot gas molecules cool in the larger space and thus lose expansive force. Having served their purpose, spent combustion products must be expelled on the next stroke to make room for a fresh charge of air and fuel.

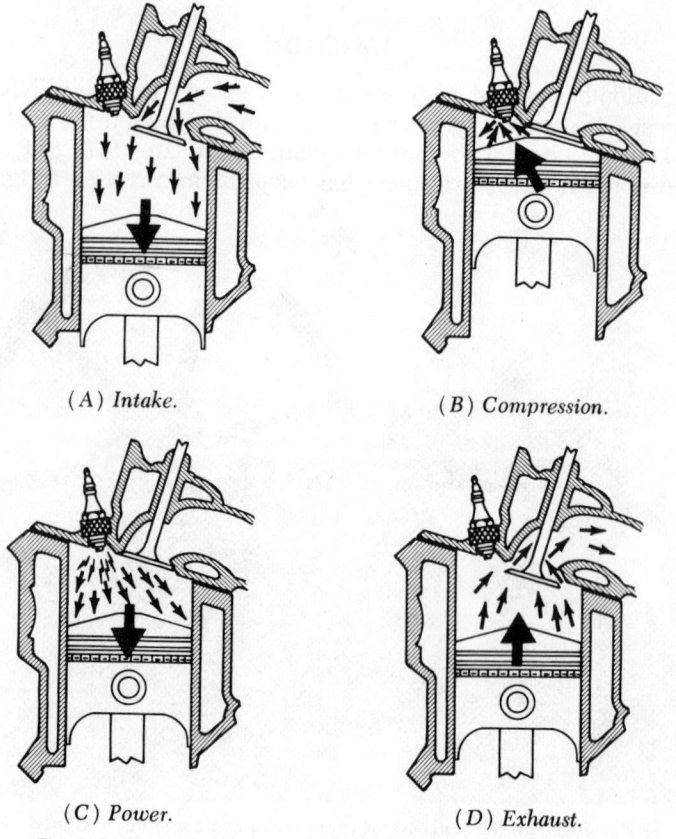

(A) Intake.   (B) Compression.

(C) Power.   (D) Exhaust.

Fig. 1-2. The four strokes of the modern Otto-cycle engine.

Intake, compression, power, and *exhaust* (Fig. 1-2)—the four strokes which constitute the *Otto cycle*—have been endlessly enlarged upon in automotive engine designs. Though the shapes and arrangement of parts may vary, the phases of the Otto cycle are equally recognizable in both the time-honored piston engine and the newer Wankel rotary engine. Differences of appearance aside, however, *all* Otto-cycle engines share a common critical need for reliable, powerful, and perfectly timed ignition sparks.

The critical factors in spark ignition are *timing, energy,* and *duration*. The objective is to reliably apply a precisely timed, sufficiently high potential between two electrodes to develop a high-temperature arc which will drive the fuel/air mixture well into its kindling temperature range. The difficulties of consistently obtaining this objective are easy to see when you consider that an eight-cylinder engine turning over at 3000 rpm requires about 12,000 precisely timed and dimensioned sparks per minute.

Not surprisingly, the mechanical and electrical demands upon an ignition system are very great. The generation and delivery of spark potential must be carefully synchronized to the mechanical movement of engine parts over a widely varying range of speeds and loads. The ignition system must unfailingly deliver enough spark energy at exactly the right instant to each spark plug. A misfire in just one cylinder out of eight noticeably reduces performance, but also raises exhaust emissions tenfold and sharply increases operating and maintenance costs. Therefore, as a prelude to our discussion of ignition systems, let us examine the basic nature of the spark which ignites the fuel/air mixture.

THE SPARK PLUG AND ENERGY CONVERSION

It is the function of a spark plug to provide concentrated conversion of electrical energy into thermal energy at a location suitable for fuel ignition. Sufficient heat energy must be dissipated in the plug gap to kindle the fuel charge within the gap and commence combustion. This fascinating electrical-to-thermal energy conversion at the plug gap and the resultant elevation of the spark plug to the status of a *transducer* deserve closer study.

What is commonly called a *spark* is but another name for a highly ionized luminescent column of high-temperature gases. Although references to "big fat blue sparks" and "little skinny red sparks" are common in ignition literature, the fundamental qualities of the spark are almost never discussed. Let's correct the situation.

The parameters of a spark include its physical dimensions and the thermal values of the high-temperature ionized gas column which comprises the spark. This involves the diameter and length of the

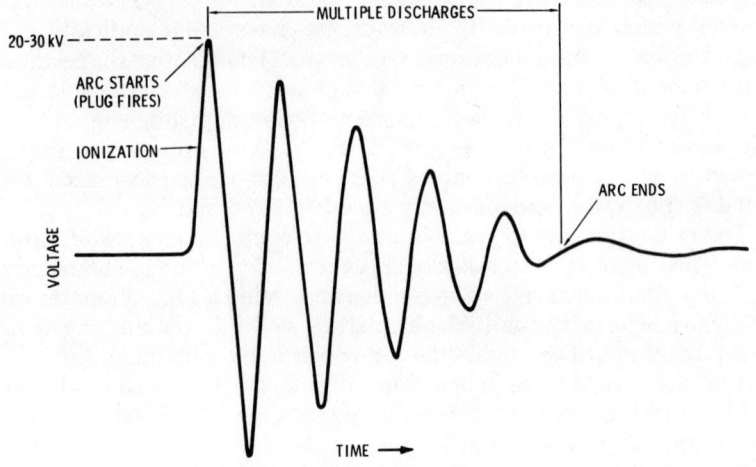

(A) Spark-voltage waveform.

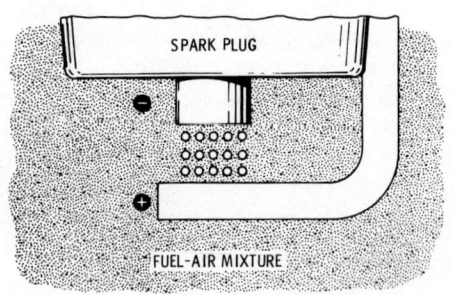

(B) Ionization.

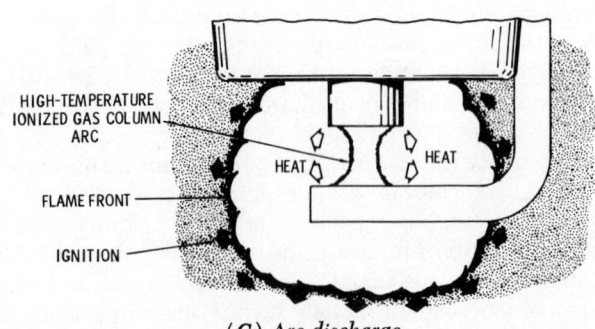

(C) Arc discharge.

Fig. 1-3. Electrical-to-thermal energy conversion within the spark-plug gap.

column, the specific heat of the gas, and the core temperature of the gas column, which can reach 4000°C.

In the first stage of ignition, a steeply rising voltage is impressed across the plug electrodes, ionizing the mixture as shown in Fig. 1-3. At some combination of high voltage and decreasing gap resistance, electrons flow between the two electrodes—slowly at first, then in avalanche fashion. Electrical-to-thermal energy conversion then occurs within the spark plug gap in two distinct phases: first, the electrical discharge (spark) generates a high-temperature gas column (arc); second, in the thermal phase, energy radiates from the arc into the surrounding fuel/air mixture. (Actually, the duration of the high-temperature gas column (arc) is a *thermodynamic phenomenon*, not directly associated with the electrical discharge which produced it.) Very high localized temperatures are created which dissipate into the fuel/air mixture, raising it to its kindling temperature. A flame front advances outward from the gap into the remaining volume of fuel/air mixture within the combustion chamber.

In terms of duration, a short, high-energy discharge can have the same ignition effectiveness as a longer, lower-energy discharge. However, mechanical factors assume greater prominence in how well a given charge of fuel can be burned. The flame front which proceeds outward from the arc produced at the plug gap must be correctly channeled by the combustion chamber, so as to reach and ignite the largest volume of fuel in the critical thousandths of a second preceding the power stroke. It is important to realize that most combustion chambers are typically designed around an idealized model of a conventional ignition system, and that the engine designer has assumed not only a certain energy discharge at the plug gap but also a typical duration. Obviously, a change made in the ignition system which increases energy but abbreviates duration will have a negative influence on total burn time and, therefore, can increase the engine's emissions. Conversely, a system change which extends the duration of the ignition arc may improve combustion but it will also hasten deterioration of the spark plugs, due to the excessive heating to which the electrodes are subjected.

While these are secondary points, compared to the major problems of "misfire," they do tend to illustrate the careful matching necessary between mechanical design of the engine and ignition-system design.

CHAPTER 2

Conventional Ignition

KETTERING IGNITION

Charles Franklin Kettering's ignition system, which he designed in 1908, is shown in his own hand in Fig. 2-1. The modern version, shown in Fig. 2-2, is regarded as *conventional ignition*. The design of the conventional ignition system relies on the storage of energy in a closed inductive circuit. Battery current flows through closed electrical contacts (breaker points) into a primary coil consisting of relatively few turns of heavy wire. The current flow establishes a magnetic field around the primary coil and also around a coupled secondary coil consisting of hundreds of turns of fine wire. When the contacts are opened, the battery current flow is interrupted and the magnetic field collapses, cutting through the coil windings and inducing a high voltage across the secondary coil. Connecting the ends of the secondary coil across a spark plug gap of finite width results in a spark discharge between the electrodes of the spark plug, since the induced voltage is proportioned to be great enough to ionize the intervening mixture. This provides an arc dissipating enough heat energy to start combustion and thus fire the fuel/air charge within the enclosed cylinder. Construction of a modern ignition coil is shown in Fig. 2-3. The advance mechanisms shown in Fig. 2-2 are there to assure that, under every condition of engine operation, ignition takes place at the most appropriate time. The factors which determine the most favorable ignition time are engine power demands, fuel economy, and decontamination of exhaust gases. The advance mechanisms shift the instant of ignition to the

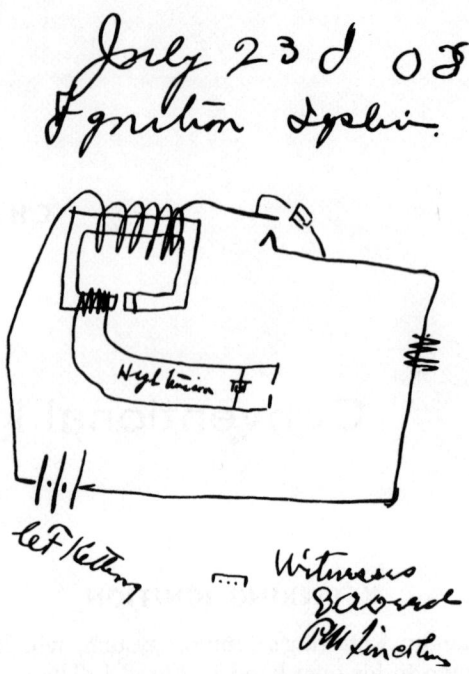

Courtesy Delco-Remy Division, General Motors Corporation

Fig. 2-1. Reproduction of Kettering's patent drawing of the original battery ignition system.

most advantageous point in time as determined by these factors. In no way, however, do the advance systems affect spark energy and duration.

PROBLEMS OF CONVENTIONAL KETTERING IGNITION

The breaker-point set, inherited from Charles Kettering's original battery ignition system, has proved to be a source of endless ignition problems throughout the more than half a century in which it has been in use. The basic problem of the conventional ignition system is caused by the acceptance of a *compromise* between reasonable breaker-point life and the maximum current the points can handle. Reasonable point life can only be expected when the maximum current through the breaker points is maintained under 4 amperes. Fig. 2-4 illustrates the accelerated wear-out curve of breaker-point sets subjected to higher current. The curves shown in Fig. 2-5 confirm the prolonged life of point sets exposed to lower break currents.

14

Automotive ignition theory has established that at least 30 millijoules of energy must be stored in the field flux of the ignition-coil primary, and then released each time the points open. The energy stored in the flux of the ignition-coil primary is related to the inductance of the primary in such a way that, allowing 4 amperes maximum current, the turns ratio of the coil in a conventional ignition

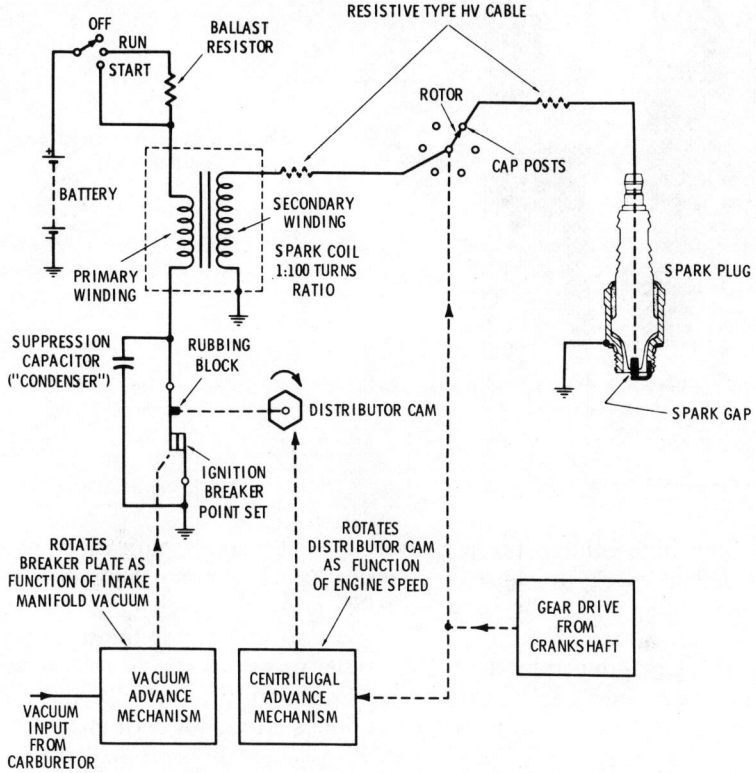

Fig. 2-2. Functional drawing of conventional Kettering ignition system.

system cannot exceed 100:1. Even this ratio is a weighted compromise since the primary inductance of a coil with a 100:1 turns ratio prohibits the storage of 30 millijoules of energy when engine speed rises. In fact, the available high voltage at the secondary of a conventional ignition coil drops to about 63% of the desired level when the engine is turning over at 4000 rpm (see Fig. 2-6). This is because the primary inductance of a conventional ignition coil and the tendency of rapidly making-and-breaking points to "bounce" do not permit enough time for 30 millijoules of energy to be stored within the primary of the coil.

15

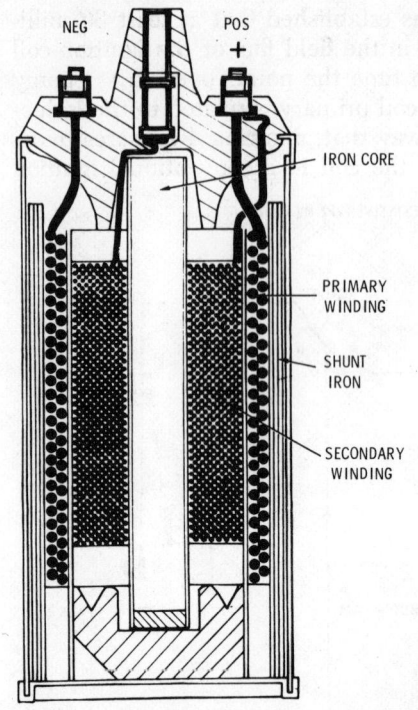

Fig. 2-3. Construction of an ignition coil.

The high-voltage pulse delivered to the spark plug must have sufficient energy to cause an electrical spark to jump the gap of the spark plug (normally between 0.025 and 0.040 inch) and fire the fuel/air mixture. Since the system must also overcome the resistance at this gap induced by the high cylinder pressures (up to 200 pounds per square inch) and a varying fuel/air mixture, a conventional Kettering ignition system is under enormous stress. Modern engine de-

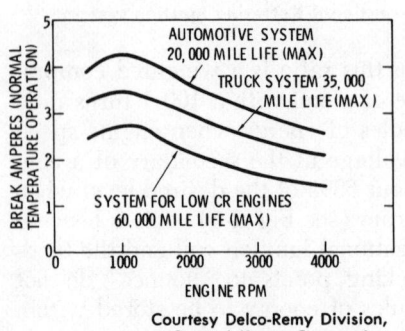

Fig. 2-4. Effects of break current on point life.

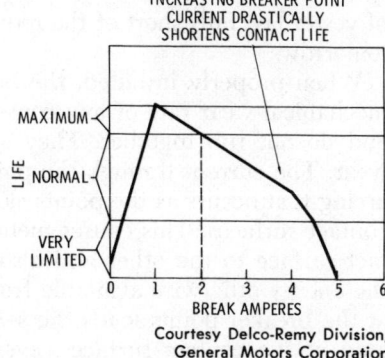

Fig. 2-5. Point life for various applications at different break currents.

signs demand a consistent spark-plug voltage in the neighborhood of 25,000 volts throughout all speed ranges for reasonable performance and economy. Less than perfect operation of a Kettering ignition system means misfiring, fouled plugs, high exhaust emissions, and wasted fuel.

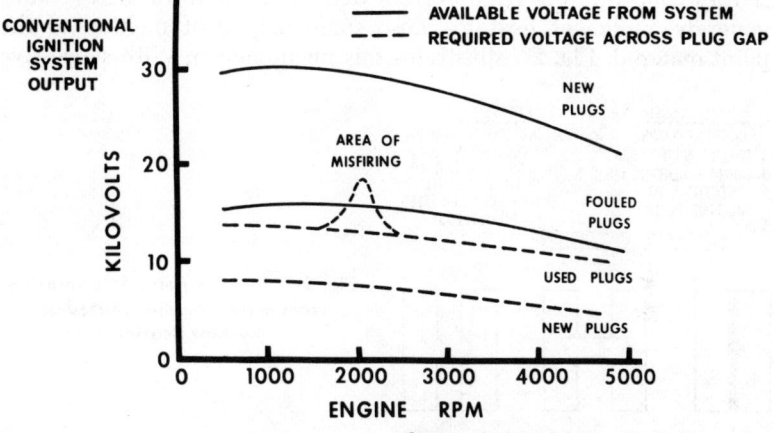

Fig. 2-6. Comparison of available and required spark-plug voltages in a conventional ignition systems.

In the conventional Kettering ignition system, the breaker points cannot remain closed long enough to store sufficient energy in the ignition coil so that adequate firing voltages can be reliably attained at turnpike speeds. Unfortunately, the point contact area cannot be made larger without the mechanism becoming cumbersome and unworkable. Thus, the compromise between point contact area and current-carrying ability produces a delicate balance between point life and engine performance that, although reasonable in the cars of yesterday, falls short of the requirement for the cars of today and tomorrow.

When properly installed, the breaker points are not subject to a mechanical-wear rate of any consequence, since they meet squarely and do not rub together. They are, however, subject to electrical wear. The current through the points when they are closed and the arcing that occurs as the points slowly open tend to erode the metal contact surfaces. This causes metal to be transferred from one contact surface to the other and also results in dissipation of some of the energy otherwise available from the secondary, since the arcing at the breaker points loads the secondary. The extreme heat generated at the contact surfaces, even when driving only limited distances at moderate speeds, is a chief enemy of point life. No matter how smooth the point surfaces appear to the naked eye, when magnified, they will always show "hills and valleys" in the metal surfaces. When two such surfaces meet, current flows only through those areas that are in contact. If two "hills" are in contact, only a tiny area of the total point surface is conducting and the distribution of current is very concentrated. High current density through a low resistance generates sufficient heat to cause some degree of melting of the point material. Fig. 2-7 illustrates this phenomenon. With successive

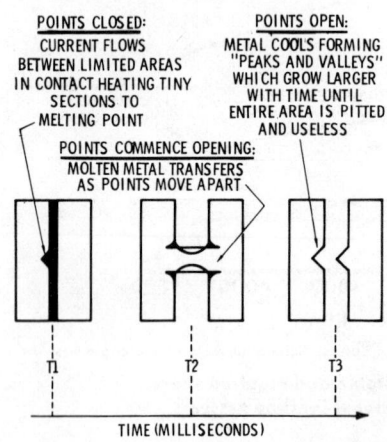

Fig. 2-7. Breaker-point deterioration from metal transfer caused by current heating.

openings and closings, isolated melting of the point contacts produces a mottled gray appearance in place of the shiny brilliance of a new point surface. With prolonged use, the mottling of the point surface gradually turns to pitting, and definite wear patterns emerge.

The electrical effect of this gradual metal transfer is to increase the resistance of the points when closed, so that the maximum current that can flow is substantially below that of a new contact set. As points age and their resistance increases, less current flows through the primary of the ignition coil, so that the required 30-millijoule stored energy level is not achieved, even at relatively low engine speed. Hence, the coil is incapable of delivering the required high-voltage spark to the plug when the aging breaker points open. This is the principal reason for deteriorating performance and starting failures in the conventional ignition system. Reliance upon mechanical breaker-point contacts limits optimum operation of the conventional system to not more than 200 hours before parts replacement becomes essential to arrest deterioration in ignition-system performance and emission control.

CHAPTER 3

A Practical Review of Semiconductor Theory

Compared to the rather crude action of mechanically driven, metal breaker-point contacts, semiconductors can approach *ideal switches*. Theoretically, an ideal switch is a device which has no power loss when continuously connecting and interrupting current flow through a load, and which is able to change from the *on* state to the *off* state in zero time. This simple definition has some far-reaching implications. It requires, for example, that in the *on* state, an ideal switch must have *no voltage drop* across its terminals. In the *off* state, the ideal switch must *cut off all current flow* through the load. And finally, the cycling speed of the ideal switch must be *infinite*, corresponding to zero time delay between one state and the other. A switch that meets all these criteria would surely be an ideal replacement for mechanical breaker points. Unfortunately, such an ideal switching device does not exist. However, semiconductors, though they fall short of this perfection, represent the best available components for switching in automotive ignition applications. To see why, let's quickly review the characteristics of semiconductors.

THE SEMICONDUCTOR JUNCTION

Germanium and silicon are metallic crystalline elements which can be chemically doped by infusion of minute quantities of impurities to create two electrically dissimilar forms of the original element. These two forms are called *p-type* and *n-type*. The p-type form of either germanium or silicon is prepared to freely accept electrons which enter into its structure. Conversely, the n-type form purposely has an oversupply of electrons in its structure.

If a p-type form is alloyed with an n-type form, a *pn junction device* is created as shown in Fig. 3-1. The connection of a wire to each end of the junction device produces a *diode*.

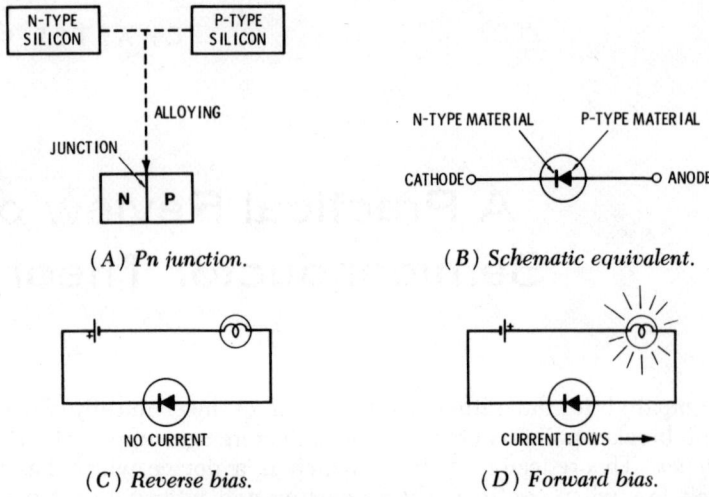

Fig. 3-1. Semiconductor diode structure.

THE SEMICONDUCTOR DIODE

The most useful characteristic of a diode is that it allows current to pass in only one direction. The *anode* (p-type half of the junction) must be positive with respect to the cathode (n-type half of the junction) in order for the device to pass a current and thus behave like a closed switch. This condition is called *forward bias*.

Only a relatively small amount of forward-bias voltage (anode positive) is required to produce a forward current which increases rapidly as the anode voltage increases slightly. On the other hand, only a very slight amount of current flows when the anode is negative. Increasing the negative voltage on the anode produces a very small increase in current until the peak reverse voltage (prv) is reached. The forward and reverse characteristics of a diode are shown in Fig. 3-2.

All diodes have a voltage drop when conducting a forward (normal) current. The forward voltage drop (V_F) for a silicon diode is about 0.7 volt, and for a germanium diode, it is about 0.3 volt.

The current rating of a diode is determined entirely by the internal temperature which it can withstand. High-current semiconductor devices must be fastened to a metal plate called a *heat sink*. The heat sink conducts heat away from the device and enables the silicon pellet to remain relatively cool. A low-current device usually relies on the air circulating around its case for cooling.

One of the disadvantages of diodes is that they cannot withstand infinite reverse voltage. Just as the air gap between switch contacts will break down if the voltage across them is high enough, the blocking action of the diode junction will also break down if the peak reverse voltage is exceeded. The voltage at which this occurs is

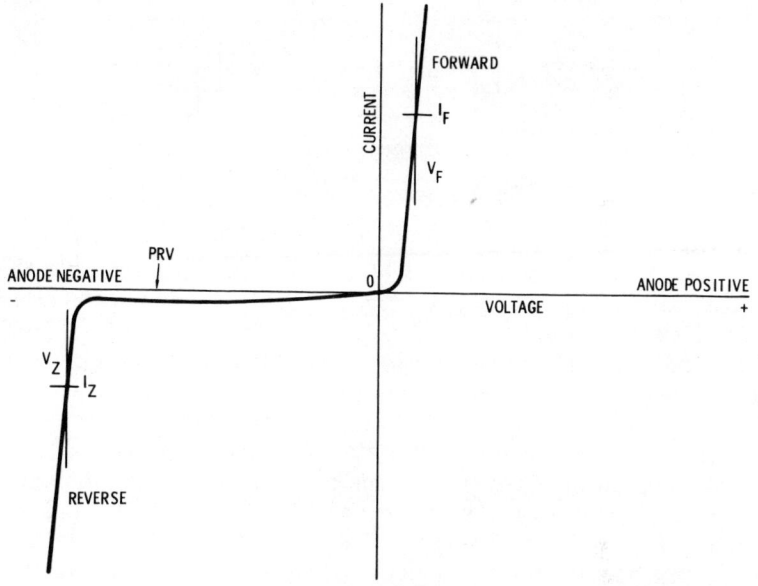

Fig. 3-2. Characteristics of a diode.

called the breakdown or *zener* voltage (V_Z). In effect, as a certain critical zener voltage is exceeded, the diode will conduct in the reverse direction. If no current limiting is provided, I_Z rises sharply with increasing voltage and the junction heats excessively. If this is carried to extremes, the diode junction will be destroyed.

Fortunately, there are benefits in this apparent weakness. The *zener diode* is a device carefully designed to break down at a specific reverse voltage. Such a diode has a nearly constant voltage drop for any current within its operating range. This feature makes it a good *voltage-regulator* or *voltage-reference* device. Current through the zener diode must be limited by external resistance. Also, it must be remembered that in the forward-bias direction the zener diode behaves like an ordinary diode.

23

THE TRANSISTOR

Suppose that instead of forming a single pn junction, *two* junctions are formed in a sandwich, using p-type material on either side of a single n-type slab, as shown in Fig. 3-3. Such a device would be called a *pnp junction transistor*. If you used n-type material on either side of a common p-type slab, you would have an *npn junction transistor*. Both structures are commonly available in silicon and germanium transistors.

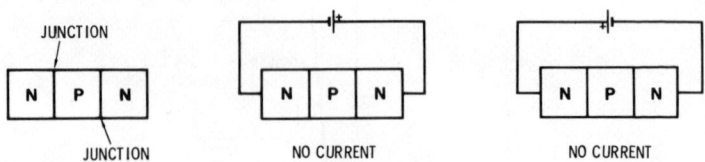

Fig. 3-3. Transistor structure.

In theory, the formation of a two-junction device prevents current flow between the two ends of the sandwich, regardless of the polarity of the applied voltage (Fig. 3-3). One junction or the other will block the current flow. However, if a wire is brought out from the middle slab of the sandwich (Fig. 3-4), the situation changes considerably. Now it is possible to bias the transistor so that current flow becomes possible. For convenience, we label the lead wires from the transistor structure: *emitter, base,* and *collector*.

Assuming an npn structure, if one of the n-type slabs (the emitter) is connected to the negative pole of a battery (B1) and the center p-type slab (the base) is connected to the positive pole, that diode junction becomes conductive because it is forward biased. Suppose now that we connect a second battery (B2) between the center p-type slab and the opposite n-type slab (the collector) so that the junction between them is reverse biased. Interestingly enough, as current from battery B1 flows through the first junction (between emitter and base), electrons enter into a region of the

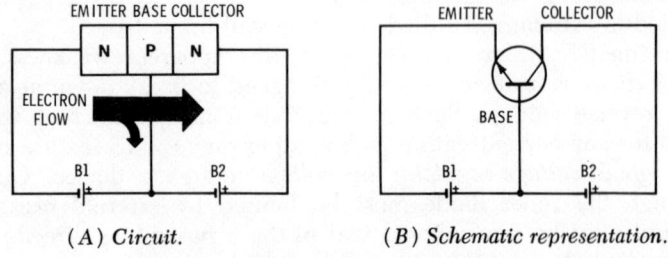

(A) *Circuit.*   (B) *Schematic representation.*

Fig. 3-4. Forward-biased npn transistor.

24

p-type slab where they can be attracted across the second junction (between base and collector) by the strong positive collector potential due to battery B2. In practical devices, 95 to 99.5% of the current entering the first junction reaches the collector region. This high percentage of current penetration provides power gain in the output circuit and is the basis for the ability of a transistor to amplify and act as a switch. In effect, the flow of a relatively small control current through the emitter-base junction governs the flow of a larger current between the emitter and collector. When the level of the control current is changed, the larger collector current changes too. If the control current stops, the collector current ceases also.

A forward-biased pnp transistor is shown in Fig. 3-5. Again, as for the npn transistor, the emitter-base junction is forward biased by battery B1. However, in this case, current flow is from the base to the emitter. Electrons near the collector-base junction of the n-type base region are attracted to the emitter. Since there is a high negative potential on the collector due to battery B2 and a depletion of electrons on the base side of the junction, electrons move across

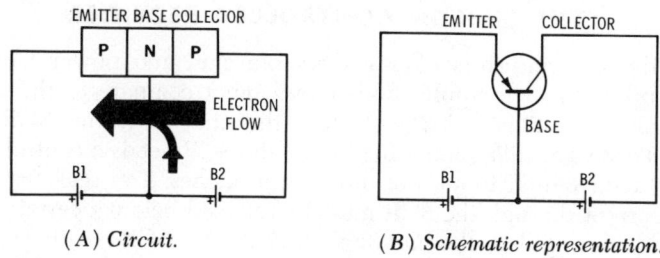

Fig. 3-5. Forward-biased pnp transistor.

the collector-base junction into the base region where they are immediately attracted to the emitter. Due to the high negative potential on the collector, it furnishes many more electrons to the emitter than does the base. Just as in the case of the npn transistor, a relatively small current through the emitter-base junction controls a much larger current flow from the collector to the emitter. This control action of the base enables the transistor to be used as an amplifier or as a switch.

Because transistors are made up of diode-like junctions, they exhibit similar characteristics in many respects. For example, the junction between collector and base is a reverse-biased diode and has a definable zener-point voltage (V_{CBO}). Accordingly, if the collector-to-base voltage exceeds that level, the reverse-breakdown current flow through the device may damage the transistor. The same effect can occur between the collector and the emitter, where the breakdown voltage is called the V_{CEO}. It is important that the transistor be operated within the range of the manufacturer's published ratings. Just as important is the fact that the heating of a transistor junction can allow small currents to leak past a normally blocking interface. When this occurs in the emitter-base junction, bias conditions shift and the operating characteristics of the transistor change. Most well-designed transistor circuits include components which stabilize bias and minimize the effects of leakage currents.

The phenomenon of transistor operation works at high currents of many amperes as well as at levels of a few milliamperes. But, since a transistor is not a perfect switch, some internal heating occurs when current passes through the transistor junction. This heat must be dissipated, or the junction may be damaged or destroyed. Transistors designed to handle heavy currents are physically large structures. They are packaged in heavy metal cases or provided with heat-conducting tabs, so that thermal energy built up in the junctions can be transferred to a larger mass (a heat sink) and dissipated in the surrounding air.

THE SILICON CONTROLLED RECTIFIER

Although transistors offer respectable gain and power handling capability, in some applications they cannot compare to the silicon controlled rectifier (SCR). Unlike the transistor, the SCR is a *regenerative switch*. Operating on dc, the SCR needs a control pulse to switch from off to on; but once on, it latches. To break the latch, the current through the SCR must be reduced below a certain minimum value called the *holding current*. An SCR switches more quickly from one state to another than does a transistor, and is capable of handling higher energy levels. The reason is simply that an SCR has enormously greater *gain*. It passes from off to on so rapidly (generally in a microsecond or two) that it offers virtually no intermediate resistance to current flow. Hence, the SCR dissipates much less internal heat than a transistor does when switching current to a load.

The high gain of an SCR is obtained by forming a four-layer (pnpn) junction device (Fig. 3-6). As in the transistor, no current can flow when a potential of either polarity is connected between anode and cathode of the SCR. This is the so-called "blocking state." Let us now, however, separate the four layers into two discrete sections as shown in Fig. 3-6C. The result is a pnp structure joined to an npn structure. In other words, the equivalent of an SCR is a regeneratively connected pair of transistors; one pnp and one npn. If you study the two-transistor equivalent circuit of an SCR in Fig. 3-6D, you will see what makes the blocking state stable. Although there is positive potential at the emitter of Q1 (SCR anode) and a negative potential at the emitter of Q2 (SCR cathode), both transistors are reverse-biased and nonconducting. However, if a positive pulse of current is applied to the base of Q2 (SCR gate), the emitter-base junction of that transistor is momentarily biased on. Instantly, Q2 conducts, causing the base-emitter junction of Q1 to be forward biased. The collector of Q1 also sees a return path through the forward-biased base-emitter junction of Q2. The instant that Q1 turns on, Q2 receives sufficient forward-bias current that it immediately reaches saturation. In so doing, it increases the base-emitter current of Q1 so that it, too, becomes saturated. Instantly, both transistors are fully turned on. They remain in that state until the dc current flow through both is reduced below the level at which their junctions can remain forward biased. For dc operation, this generally means either opening the circuit, momentarily shunting the current around the junctions, or reducing the voltage applied between anode and cathode to such a low level that the holding current cannot be sustained and the device blocks.

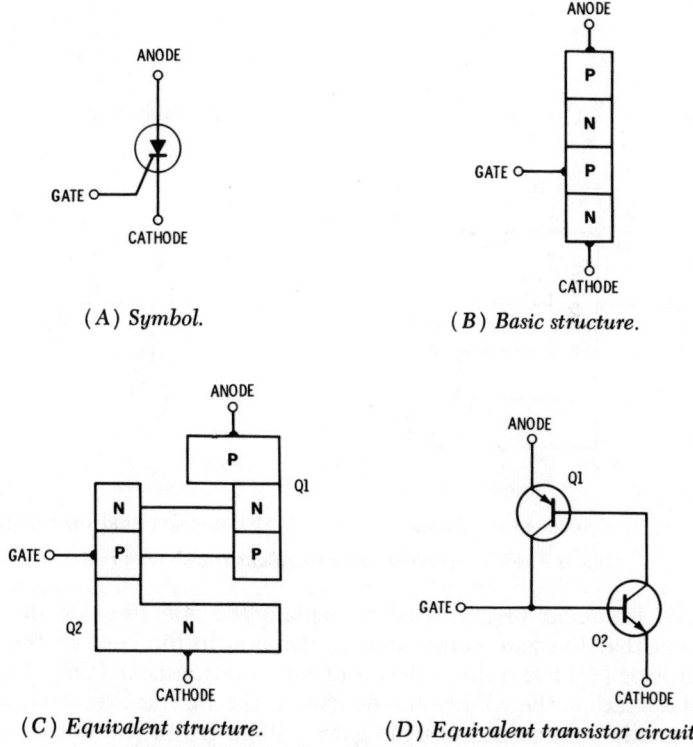

(A) Symbol. (B) Basic structure.

(C) Equivalent structure. (D) Equivalent transistor circuit.

Fig. 3-6. SCR structure and equivalent transistor circuit.

THE PROGRAMMABLE UNIJUNCTION TRANSISTOR

Similar in design to the SCR, the structure of the programmable unijunction transistor (PUT) is also based on the use of four alternate layers of p-type and n-type material (Fig. 3-7). A marked difference is that the position of the gate is changed, creating a "complementary SCR" in which the gate is related to the anode, rather than the cathode. This means that the device is turned on when the gate is *negative* with respect to the anode.

As shown in Fig. 3-7D, the PUT has a two-transistor analogy

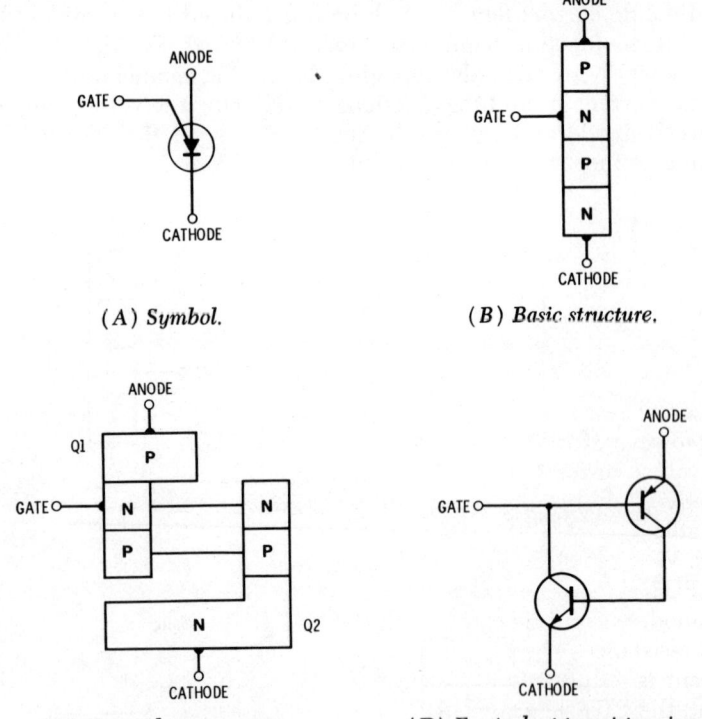

(A) *Symbol.* (B) *Basic structure.*

(C) *Equivalent structure.* (D) *Equivalent transistor circuit.*

Fig. 3-7. PUT structure and equivalent transistor circuit.

which is similar to that used to explain the operation of an SCR, except that the gate connection is common to the base of the pnp transistor (Q1) and the collector of the npn transistor (Q2). In normal operation, the voltage on the gate of the PUT is fixed and, when the anode voltage exceeds the gate voltage by an amount equal to the voltage drop across the base-emitter junction of Q1, the anode-to-cathode resistance is reduced. The anode voltage necessary to start conduction is called the peak voltage (V_P).

The negative-resistance characteristic for a PUT with a fixed gate voltage is shown in Fig. 3-8. For anode voltages (V_A) less than the peak voltage (V_P), the PUT exhibits a positive incremental resistance characteristic (region 1). Also, for anode currents above the valley current (I_V) which occurs at the valley voltage (V_V), the in-

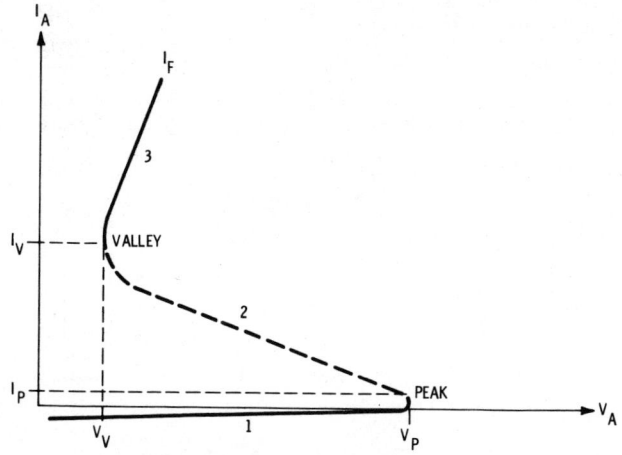

Fig. 3-8. Characteristics of a PUT.

cremental resistance characteristics of the PUT are positive (region 3). However, for anode currents between the peak current (I_P) and the valley current (I_V), the incremental resistance characteristic is negative. This is region 2 on the characteristic curve. This region is unstable and forms the basis for use of the PUT in oscillator circuits. When the anode voltage (V_A) exceeds the peak voltage (V_P), the PUT will regeneratively switch to its low-resistance state and the anode current increases rapidly to a level limited by the external load resistance. The PUT will remain in the on state until the anode current is reduced to a level below the valley current (I_V). At this point, the PUT returns to its blocking or off state.

CHAPTER 4

Understanding Solid-State Ignition Systems

All ignition systems presently in use or contemplated in the future can be grouped into two main branches as shown in Fig. 4-1. These two main branches are *the inductive-discharge systems* and the *capacitive-discharge systems*. The inductive-discharge systems have the longest history and include the conventional Kettering system and the newer magnetically triggered transistor switching systems. The conventional Kettering system has three subbranches of solid-state ignition systems: the *contact-triggered* transistor switching systems, the *contact-isolator* switching system, and the *contact-assistance* dwell-extender system. The capacitive-discharge systems includes only two subbranches: the *inverter-type* system and the *magneto-type* system. Since the objective of any ignition system is to create a "good" spark, this separation of systems into family groups is principally important in sorting out the methods and techniques used in the various systems.

INDUCTIVE-DISCHARGE SYSTEMS

When current flows through an inductor (spark coil) and establishes magnetic flux about the inductor turns, energy is said to be *stored* in the inductor. With the abrupt cessation of current flow, this flux collapses and, in cutting through the inductor turns, causes the production of a large emf between the two ends of the winding. If these ends are connected to a pair of carefully spaced electrodes, the emf will originate and sustain an arc between them which will

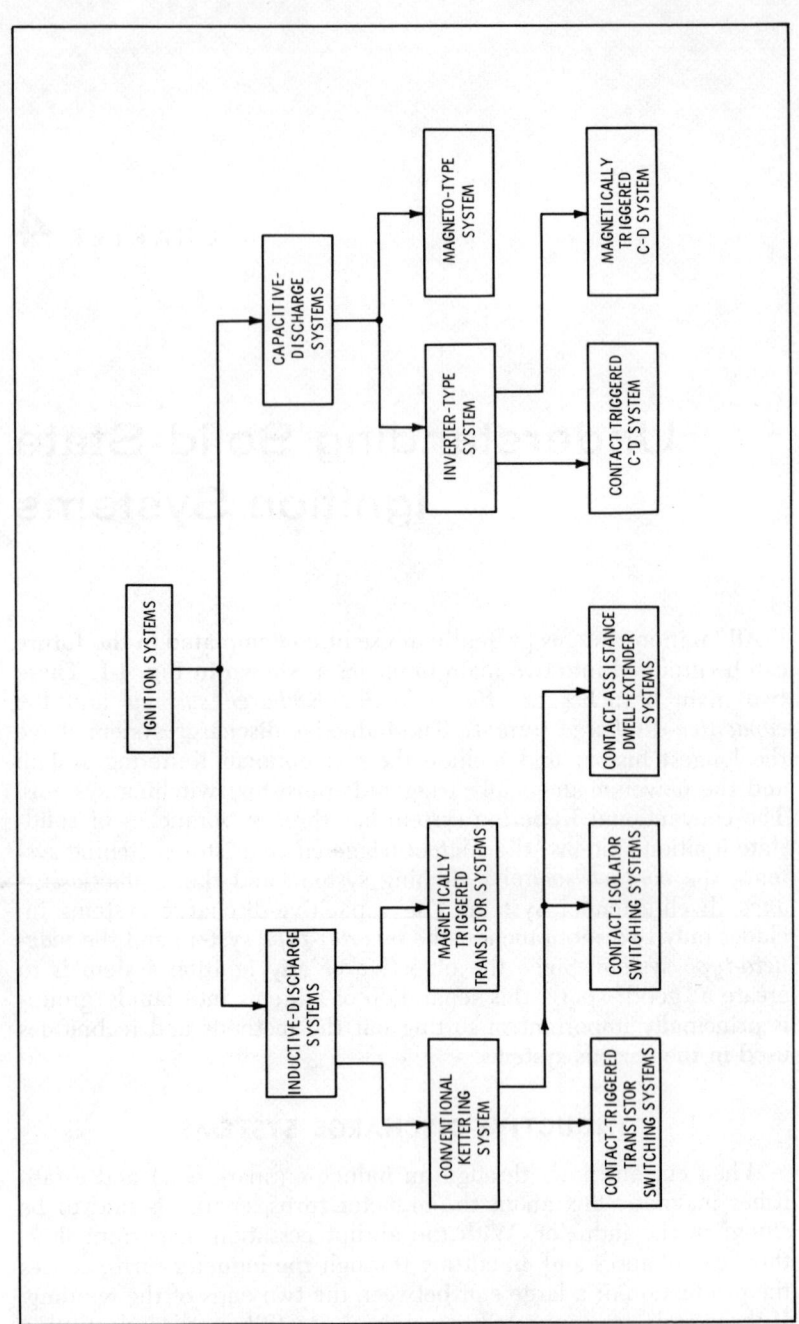

Fig. 4-1. Evolution of solid-state ignition systems.

last until the energy stored in the inductor has been discharged. The equation for determining energy stored in an inductor is:

$$e = \frac{LI^2}{2}$$

where,
e is energy in joules,
L is inductance in henries,
I is current flow through the inductor in amperes.

To store a desired level of energy in an inductor, you must use either a high-inductance (many turns) coil and moderate current or, a low-inductance coil and substantially greater current. However, a limiting factor which is not part of the energy storage equation is *time*. In automotive ignition, spark potential must appear at successive cylinders at a time determined by the speed of the engine. Thus, the time required to "pump up" the ignition system so that it is ready to deliver the proper spark energy precisely when needed must be made as brief as possible. This requires the use of a minimum practical primary inductance for the spark coil and the maximum primary current that the switching device can reliably handle.

As we learned in Chapter 2, the conventional Kettering system relies upon a coil with a moderate primary inductance (for a reasonably brief time constant), balanced against a current flow which is limited by the mechanical size and heat dissipation of the breaker points. The deficiencies of this compromise stem from the burden imposed upon the breaker points. With the deterioration of the points, the system ceases to reach required energy storage levels within the time constraints imposed by the mechanical timing of the engine.

Not surprisingly, therefore, many of the solid-state versions of the conventional Kettering system seek to take the burden off the points. This is done either by isolating the points from the coil inductance, by parallelling them to provide a supplemental path for energy flow to the coil, or by completely removing the points from the coil-charging circuit and merely using them to control a power transistor which handles the switching of current in the primary circuit of the coil. Each of these approaches has advantages and disadvantages which are individually described in later sections of the text.

A newer school of thought in inductive-discharge ignition systems contends that any system containing breaker points is prone to failure, or requires uneconomical circuitry to compensate for breaker-point deficiencies. Accordingly, many designers have devised magnetically triggered transistor switching systems which eliminate the points entirely. These magnetically triggered systems synthesize point function through use of a specialized pulse generator combined with solid-state timing and switching circuitry.

OPERATING PRINCIPLES OF SOLID-STATE INDUCTIVE-DISCHARGE SYSTEMS

In the early 1960s several bright innovators simultaneously arrived at the conclusion that the transistor could improve the performance of the automotive ignition system. The basic idea of a transistorized ignition system had been around during the 1950s, but it was not until 1963 that Ford Motor Company took note of the idea and offered a basic contact-triggered transistor ignition system as optional equipment in its 1963 heavy-duty trucks. In a similar example of gradual toe-wetting, General Motors' Pontiac Division also included a transistorized ignition system in its 1963 list of options.

The basic premise of the original transistor ignition system was to "take the load off the points." It was theorized that if a power transistor were interposed between that end of the spark-coil primary normally connected in series with the points to ground, and if the points were connected so that they merely switched bias current to the base-emitter junction of the transistor, then the transistor could accomplish the switching of the spark-coil primary current through its collector-emitter junction. In this manner, the transistor

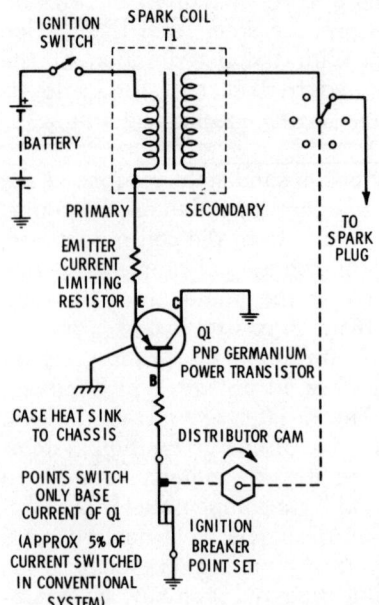

Fig. 4-2. Simplified schematic of basic contact-triggered transistor ignition system.

34

could perform the "brute" switching action, while the breaker points loafed along, switching only infinitesimal base current. Fig. 4-2 shows such an arrangement. This contact-triggered transistorized ignition system is essentially identical to the conventional Kettering system, except for the addition of transistor Q1 and the removal of the "condenser." The pnp germanium transistor has its emitter and collector leads connected in series with the primary of spark coil T1. The base lead of the transistor is connected to the distributor breaker points through a current-limiting resistor. The closing and opening of the point contacts turns the transistor on and off, which initiates and interrupts the current through the primary of the ignition coil. Sufficiently large current flows to easily store the required 30 millijoules in the primary inductance of the coil. Cessation of primary current causes the primary-coil flux to collapse, thereby generating a high secondary voltage as in the conventional system. The transistor base current through the point contacts is approximately 1/20th of the current in the primary of the ignition coil.

Now that more current could be made available to the coil primary than was possible in the conventional system, transistor ignition-system designers theorized that spark rise-time could be improved by reducing the resistance and inductance of the ignition-coil primary. Starting with the worst possible case of ignition system conditions—fouled spark plugs, high compression, rich fuel mixture, etc.—designers achieved coil designs with 250:1 and even 400:1 ratios. This redesign of the ignition coil resulted in decreased resistance and inductance of the primary to a fraction of that of the conventional ignition-coil values, halved the resistance of the secondary winding, and resulted in less stress of the transistor junctions by reducing reflected primary voltage from 250 volts at switch-off to somewhat less than 100 volts. As a final bonus, the peak primary current was increased from the 4-ampere limit of the conventional system to nearly 8 amperes. The net effect of these redesign efforts was an ignition coil with a higher rise time in which energy could be readily stored and converted into usable spark for superior, high-speed performance. Fig. 4-3 compares rise time and energy storage of conventional ignition systems and contact-triggered transistorized ignition systems.

Experimentation with contact-triggered transistorized ignition systems flourished during the 1960s. Countless transistor switching circuits were devised and published which could be easily added onto the conventional ignition systems of that era. Most of these systems could not be produced economically or suffered reliability problems. Early experience with contact-triggered transistorized ignition systems proved disappointing because the automobile represented such an inhospitable environment from the standpoints of drastic temperature extremes, harsh physical vibration, and severe voltage transients on primary supply lines. Designers quickly realized that adequate heat sinking of the transistor switch was essential, so that the junctions within the transistor could be maintained below the temperature at which the transistor would turn on by itself. Considering the typical switched current (6-8 amperes) and the wide range of temperatures under the hood, temperature stabilization proved a difficult task.

With the problems of economics and the state of the art in transistor manufacture during the 1960s, it is not surprising that automobile manufacturers saw fit to wait before plunging into transistorizing the ignition systems of their products. However, several interesting and noteworthy contact-triggered systems *did* come into existence during that era, and these are described in the following text.

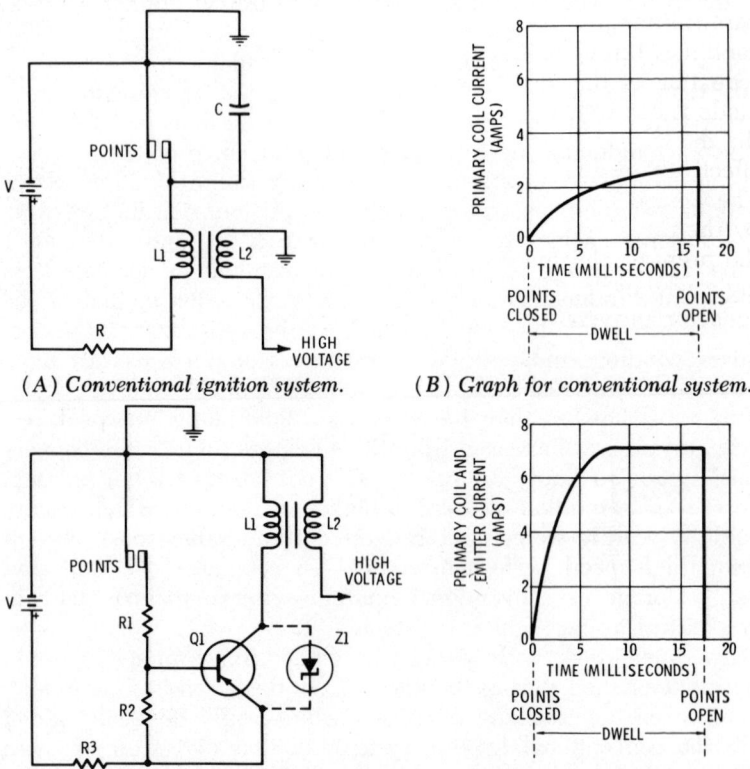

(A) Conventional ignition system. (B) Graph for conventional system.

(C) Transistorized ignition system. (D) Graph for transistorized system.

Fig. 4-3. Current rise time and energy storage in ignition-coil primary.

SINGLE-TRANSISTOR CONTACT-TRIGGERED IGNITION CIRCUITS

Fig. 4-4 shows the schematic of a typical single-transistor contact-triggered ignition system of good design. The coil has a 400:1 turns ratio to limit the reverse voltage to a value that can be safely handled by the zener diode. Since the 2N174 transistor has a collector-emitter breakdown voltage rating of only 60 volts, a zener diode is used to insure that this voltage is not exceeded.

During the "points-closed" portion of the operating cycle, resistors R2 and R3 serve as a voltage divider supplying the bias needed to turn on the transistor. Current in the primary circuit passes through the coil, through the collector-emitter circuit of the transistor, through the diode, through the ballast resistor (R1), through the ignition switch, and back to the battery. The ballast resistor serves to limit total primary current.

When the points are opened, forward bias is removed from the base-emitter junction of the transistor. A small amount of current continues through the 15-amp diode in order to provide a 0.7-volt reduction of the voltage on the emitter. This current through the diode is controlled by R4. The voltage drop thus established across the diode biases the base-emitter junction in the reverse direction, effectively increasing the breakdown voltage rating of the transistor. The voltage builds up across the primary of the coil and is multiplied by the turns ratio of the coil. At some point below 56 volts across the primary, the spark plug fires, dissipating the energy and starting a new cycle. The zener diode prevents the voltage between the collector and emitter of the transistor from exceeding 56 volts.

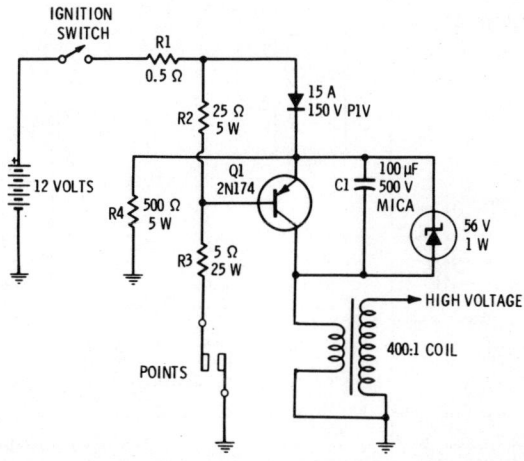

Fig. 4-4. Single-transistor, contact-triggered ignition system.

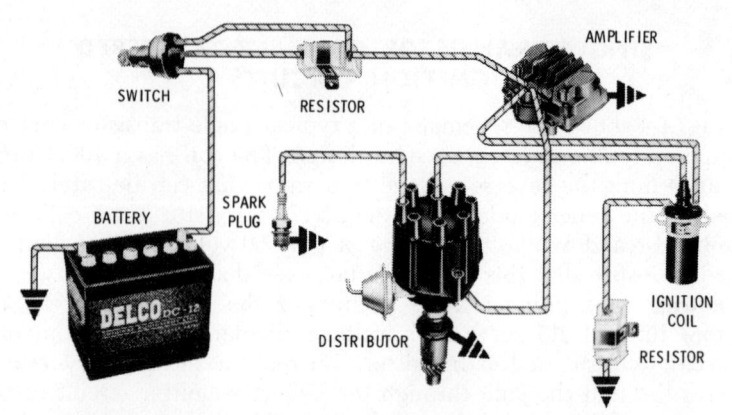

Fig. 4-5. Components of a Delco-Remy contact-triggered ignition system.

Fig. 4-5 illustrates the components of a Delco-Remy contact-triggered inductive-discharge ignition system, and Fig. 4-6 shows the circuit schematic. A pnp transistor (Q1) mounted on a cast alloy heat sink constitutes the amplifier of the circuit. The transistor is shunted by a zener diode (D1) connected between the collector and emitter. This diode supplies a path for the current to flow around the transistor junctions when high-peak positive voltages appear at the collector. If the diode were not present, the transistor would be stressed by excessive voltages each time current through the primary of the ignition coil ceases. D1 also protects the transistor should a large negative-voltage transient appear on the battery supply line. The emitter of Q1 connects to the positive terminal of the battery through a low-resistance wirewound resistor (R3) and

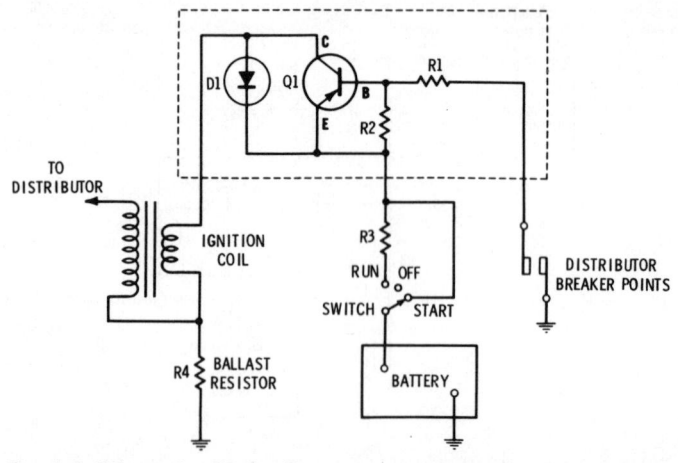

Fig. 4-6. Schematic of Delco-Remy contact-triggered ignition system.

the ignition switch contacts. In the START position, the resistor is shunted by the switch to permit maximum current flow. The resistor is reinserted into the circuit when the switch is returned to the RUN position.

Assuming that the breaker points are open, Q1 is reverse biased by R2 and no collector current flows. However, when the points close, R1 is grounded and forward-bias current is applied to the base of Q1 through the divider circuit consisting of R1 and R2. Q1 immediately switches current to the primary of the ignition coil, storing energy in the field flux so long as the points remain closed. Ballast resistor R4 limits the charging current. When the points open, the base of Q1 is reversed biased and collector current ceases abruptly. Immediately, the field surrounding the coil collapses and a high voltage is induced in the secondary as the stored magnetic energy is returned to the circuit. The high potential is delivered to the distributor rotor and thence to the spark plug through conventional TVRS (television resistance suppression) wiring.

DUAL-TRANSISTOR CONTACT-TRIGGERED IGNITION CIRCUITS

The ancestry of this design, known as the Motorola "Hot-Coil" ignition circuit, can be traced back to 1961 when it originated with Motorola engineers. This circuit remains popular with hobbyists to the present day because it delivers generally dependable performance under gruelling engine-compartment temperature extremes.

The circuit is designed to take advantage of special coil design and employs low-cost germanium power transistors. This circuit configuration makes it possible to reduce the overall cost of an ignition system that is designed around a coil with an optimum turns ratio (250:1) to the level of a system using a single transistor and a coil with a 400:1 turns ratio. To further reduce cost, 1-watt zener diodes are used in a base-bias control system that turns on the transistors should the zener voltage of the diodes be exceeded. These diodes are selected to suppress excursions past the breakdown voltage rating of the transistors being used.

In the circuit shown in Fig. 4-7, a coil with a turn ratio of 400:1 is used. Q1 and Q2 are the switching transistors for the primary current. Ballast resistor R1 limits the primary current to a safe level. This resistor must be capable of dissipating 70 to 100 watts. However, to insure stability of resistance at these high levels, R1 should be rated at 200 to 250 watts. Resistors R2 and R4 bypass collector-base leakage currents around the base-emitter junction to prevent amplification of these currents which might lead to thermal runaway and subsequent damage to the transistor.

While the points are closed, the current through the base-emitter junctions of Q1 and Q2 is amplified, causing saturation current from collector to emitter. Current rise is controlled by the inductance of the ignition-coil primary, while terminal current is a function of the amount of dc circuit resistance, including R1.

Zener diode D1 acts as a base-isolating diode in the forward-current direction and as a protective device for the Q1 collector-base circuit in the reverse, or zener, direction. The resistance introduced by D1 in the forward-current direction is compensated for by making the associated base-current limiting resistor (R5) smaller than the Q1 base-current limiting resistor (R3). Both base-current limiting resistors keep the base-emitter current levels within the design specifications for the transistor.

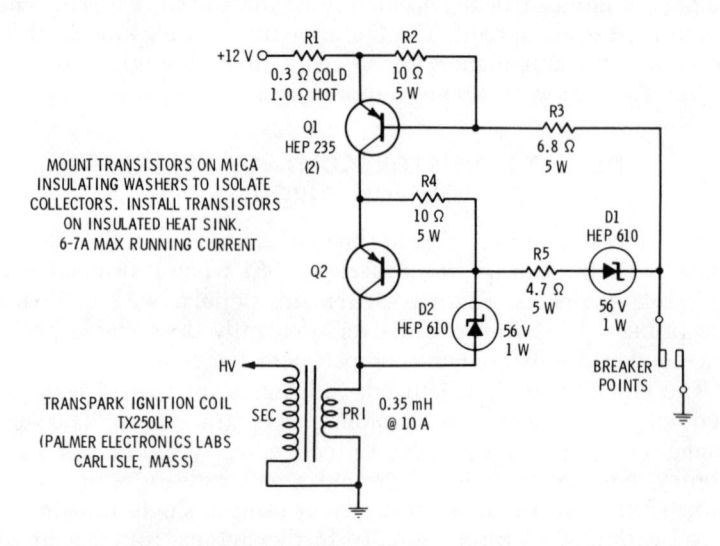

Fig. 4-7. Motorola "Hot-Coil" contact-triggered transistor ignition circuit.

When the points are opened, forward bias is removed from the bases of Q1 and Q2, and the transistors turn off. In this circuit, no provisions are made for reverse biasing the transistors. Therefore, careful attention must be paid to the collector-base breakdown voltage (BV_{CBO}) rating of the transistors to avoid exceeding it with possible consequent breakdown. The Motorola HEP devices recommended have been selected especially for this application. Should voltage levels developed in the primary of the ignition coil exceed safe design limits, diodes D1 and D2 will conduct enough current to forward bias the transistors and turn the transistors back on, damping the excessive voltage.

As shown in the schematic, the capacitor which normally shunts the breaker points in the conventional system is removed when the circuit is installed. Leaving the capacitor in would certainly reduce output and might cause transistor failure. It is recommended that the point set be replaced and set to the correct dwell angle *before* this circuit is installed. Using the specified ignition coil, spark potential in the order of 30 kV will be achievable throughout a wide range of engine speeds. Input current (engine running) should average 6.5 amperes, rising to approximately 10 amperes with the engine stalled and the points closed. Because spark-discharge duration with this circuit tends to be prolonged by virtue of the large amount of energy stored in the coil, use of spark plugs that are one heat range lower than normal is recommended. Such plugs afford better heat transfer from the electrodes, retarding gap erosion.

Simplified Dual-Transistor Contact-Triggered Ignition Circuit

Fig. 4-8 shows the schematic of a contact-triggered ignition circuit in which the collector breakdown voltage rating of the switching transistor (Q1) is increased by connecting it in series with a

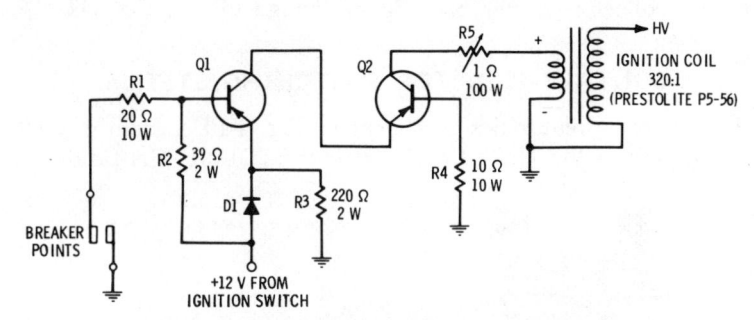

Fig. 4-8. Simplified dual-transistor ignition circuit.

like transistor (Q2). Since R4 always provides forward bias to Q2 when Q1 is switched on (breaker points closed), the sole function of Q2 is to absorb one-half of any transient voltages appearing across the primary inductance of the ignition coil at switch off. The state of Q1 is determined by the breaker points, with temperature stabilization provided by R2 and D1.

Complementary Dual-Transistor Contact-Triggered Ignition Circuit

Fig. 4-9 shows the circuit for a contact-triggered ignition system employing complementary pnp and npn direct-coupled switching transistors. The circuit operates with moderate breaker-point current and switches relatively quickly as a result of the loop gain of Q1

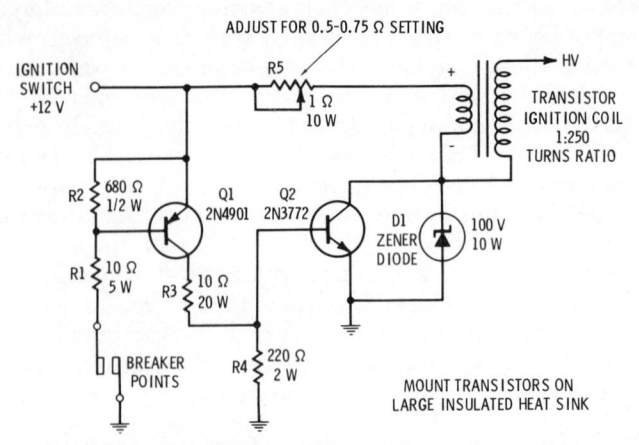

Fig. 4-9. Complementary dual-transistor ignition circuit.

and Q2. Resistors R2 and R4 provide temperature stabilization and zener diode D1 limits transients on the ignition coil primary to less than the collector-emitter breakdown voltage (BV_{CEO}) of Q2.

CONTACT-ISOLATOR SWITCHING SYSTEM

The contact-isolator design shown in Fig. 4-10 earned Mr. William H. Judson the United States Patent 3,260,891. His product is

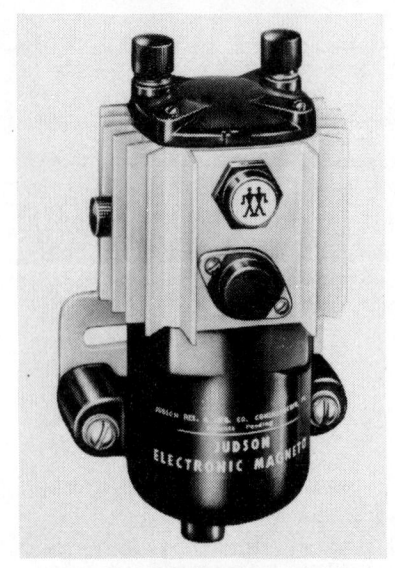

Fig. 4-10. Judson Electronic Magneto.

Courtesy Judson Research & Mfg. Co.

marketed under the trade name *Electronic Magneto*, although the circuit does not resemble a magneto in the strict sense of mechanical construction.

There are several novel features to the design. First, the system is an integrated unit in which ignition coil, semiconductors, heat sink, and other components form a neat and easily installed package. Second, despite the presence of the power transistor in the primary circuit of the ignition coil (Fig. 4-11), *all* current flows through the

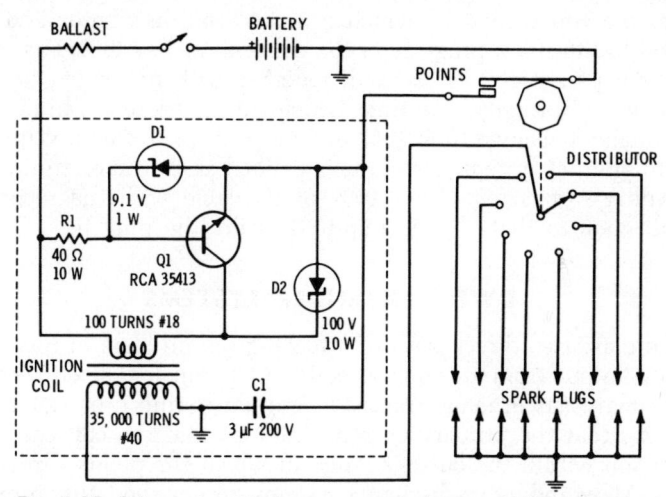

Fig. 4-11. Schematic of Judson contact-isolator switching system.

breaker points as in a conventional Kettering system. The interesting claim for this circuit, however, is that the points are effectively isolated from the primary inductance of the coil by interposing an npn silicon power transistor (Q1), shunted by a high-power zener diode (D2) in series with the primary wire to the breaker points. It is maintained that, as a result, the points switch into a resistive and capacitive load and thus are not subjected to the inductive arcing inherent in the conventional Kettering ignition system. The Judson system claims to suppress contact arcing by including a capacitor to shunt the points. This capacitor is approximately twenty times larger (3 μF) than that used in a conventional Kettering system (typically 0.15 μF).

Assuming the breaker points are closed, Q1 immediately conducts by virtue of forward bias supplied through R1 and the grounding of its emitter. Current flows through the primary of the coil rising to a claimed level of 2.5 amperes and storing energy in the primary flux. As the points open, Q1 ceases to conduct and the coil flux collapses. Voltage across the points is initially held to zero by capacitor C1 and nonconducting zener diode D2. The primary winding thus appears "open" and does not load the secondary winding, thus allowing the highest possible voltage to develop. But, as the emf induced in the primary winding exceeds the 100-volt breakdown rating of zener diode D2, any excess voltage charges capacitor C1. (When the points close again, this energy will be "dumped" through the closed contacts.) Zener diodes D1 and D2 together limit the induced primary potential which can appear between the junctions of Q1.

Meanwhile, the collapsing field flux has produced spark potential across the coil secondary which is fed to the distributor and conducted to the spark plugs as in the conventional system. Fast spark rise time (0.5 μsec.) is claimed, together with prolonged breaker-point life. (Normally, rise time is defined as the time required for the voltage to change from 10% to 90% of its peak value. With relation to ignition systems, rise time is defined as the time required for the voltage to change from 10% of its value with the secondary circuit open to that voltage required to fire the plug.)

DWELL-EXTENDER SYSTEMS

There are two direct ways of attacking the problem of increasing energy output from an ignition coil: (1) pump higher current into the primary (as in many transistor ignition circuits), or (2) extend the *time* that the primary current flows, so that the coil can reach saturation within the current range in which the points can switch safely. Many inductive-discharge systems employ the latter approach which is known as a contact-assistance *dwell extender*.

SCR Dwell Extender

The dwell-extender circuit shown in Fig. 4-12 is an interesting means of accomplishing the second approach. The idea here is to limit the period of current interruption in the primary of the coil to a tiny fraction of its value in a conventional ignition system. This means that coil-charging current flows for a much longer period than would be possible if the breaker points alone were switching the primary current. Thus, while effective dwell is increased, the ultimate result is contact assistance.

When the points are closed, the dwell-extender circuit is shorted out so that the SCR is nonconducting. During this time, the current flowing in the coil (through the closed points) builds up the magnetic field. When the points open, the emf produced by the collapsing field around the coil creates a voltage high enough to fire the spark plug. However, the instant that the points open, the positive voltage from the battery is applied directly to the anode of the SCR and, through an RC network, to its gate. About 100 micro-

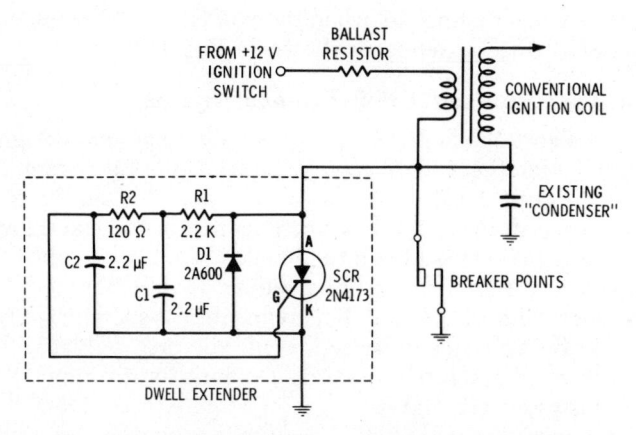

Fig. 4-12. Schematic diagram of a SCR dwell extender.

seconds after the points open, the positive voltage reaches the gate and causes the SCR to fire and latch. This closes the point circuit electrically. Shortly afterward, the points close mechanically.

The result is that the coil is being charged for almost the entire duty cycle except for the 100 microseconds needed for the spark to occur. The magnetic field built up in the coil is thus stronger and a much larger spark is available. Claims are made that the spark energy is almost doubled at high engine speeds. Diode D1 bypasses

Courtesy Radio Shack

Fig. 4-13. SCR dwell extender.

the negative spike that occurs when the points open. A commercially available SCR dwell extender is shown in Fig. 4-13.

Solitron Transistorized Dwell-Extender System

Fig. 4-14 shows the *Solitron* transistor ignition system designed to work with the ignition coil of a conventional Kettering system. Power transistor Q4 (Fig. 4-15) is biased on through R7 when the ignition switch is closed so that energy is stored in flux of the coil. However, when the engine is cranked and the breaker points open, the voltage pulse appearing across the opening points is differentiated by the network consisting of C1 and R1, momentarily switching Q1 into conduction. Because this pulse represents the first derivative of the square voltage step across the opening points, its rise time is measurable in microseconds and its duration is quite brief. Transistor Q1 is instantaneously switched into conduction by the differentiated pulse and Darlington amplifier Q2 and Q3 is forward biased. The base of Q4 is pulled down momentarily so that current is switched off in the primary of the coil. Instantly, the coil field collapses and a high-voltage pulse is induced in the secondary winding. When the differentiated pulse passes, transistor Q1, Q2, and Q3 become nonconducting and transistor Q4 is again biased on, even though the points have not closed. Q4 continues to conduct until the points have entered through another close/open sequence. The net result is an effective extension of the operative dwell period, thereby allowing sufficient time for charging the conventional 100:1-ratio ignition coil beyond the normal range of engine speeds.

Diodes D1 and D2 are used to establish a 1.2-volt minimum constant forward voltage drop when the breaker points are closed. Thus,

Fig. 4-14. Solitron transistorized dwell-extender system.

the voltage applied to capacitor C1 is not reduced to zero during the interval that the points are closed, but is held at a level which reduces the sensitivity of transistor Q1 to noise. Diode D4 prevents the base of transistor Q4 from being biased into conduction by the saturation voltage appearing across transistor Q3 during the interval that the points are open. In this way, a clean switch off of the primary current in the ignition coil occurs and a maximum spark is generated.

Switch S1 is a change-over switch which allows the choice of either transistor or conventional ignition. The switch may be used to quickly convert the system to standard Kettering-type ignition for tune-up purposes or if a failure occurs in the electronics.

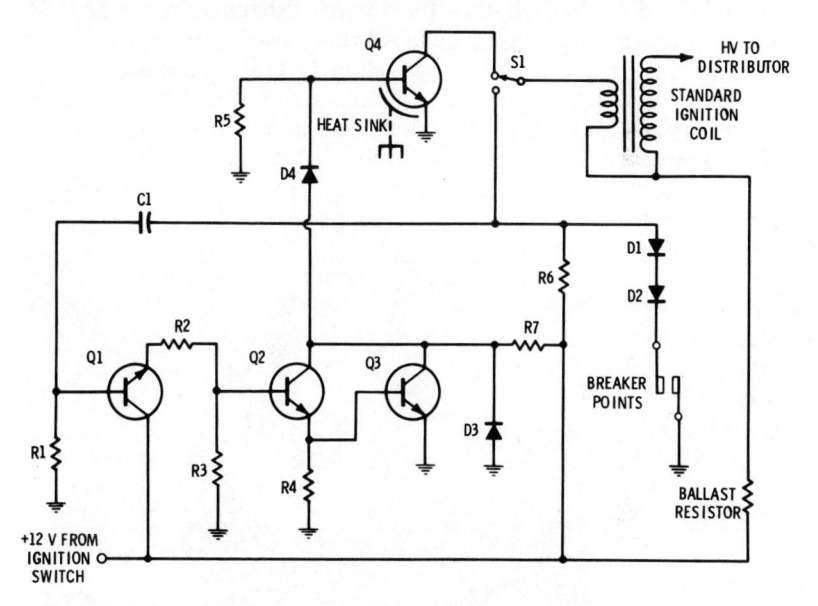

Fig. 4-15. Schematic of Solitron transistorized dwell-extender ignition system.

MAGNETICALLY TRIGGERED INDUCTIVE-DISCHARGE IGNITION SYSTEMS

Despite the advances in spark performance made possible by previously described solid-state ignition circuits, the existence of "make-break" contacts still poses the possibility of sudden ignition failure. Consider the gross physical motion of a many-sided steel cam driving a fiber or nylon rubbing block. Even if it is properly lubricated, some wear is inevitable, resulting in a timing change with mileage. If the block wears down, the points may not ever

47

open. And that spells *failure,* even though the point current has been reduced to a minimum! Accordingly, the next advance in inductive-discharge ignition has come with the elimination of the breaker points through introduction of *magnetically triggered* transistor ignition.

Delco-Remy Magnetically Triggered Transistor Ignition System

Elimination of the breaker points in solid-state ignition systems began in 1965 when the Pontiac Division of General Motors Corporation introduced a Delco-Remy magnetically triggered transistor ignition system as a special option in its *Tempest* line. The system was achieved by redesigning the entire distributor assembly. Although the external appearance of the distributor resembles a standard unit, the internal construction is quite different. As can be seen in Fig. 4-16, the breaker points are replaced by a rotating iron timer core and a magnetic pickup assembly. The timer core replaces the cam used in the conventional distributor. This core has

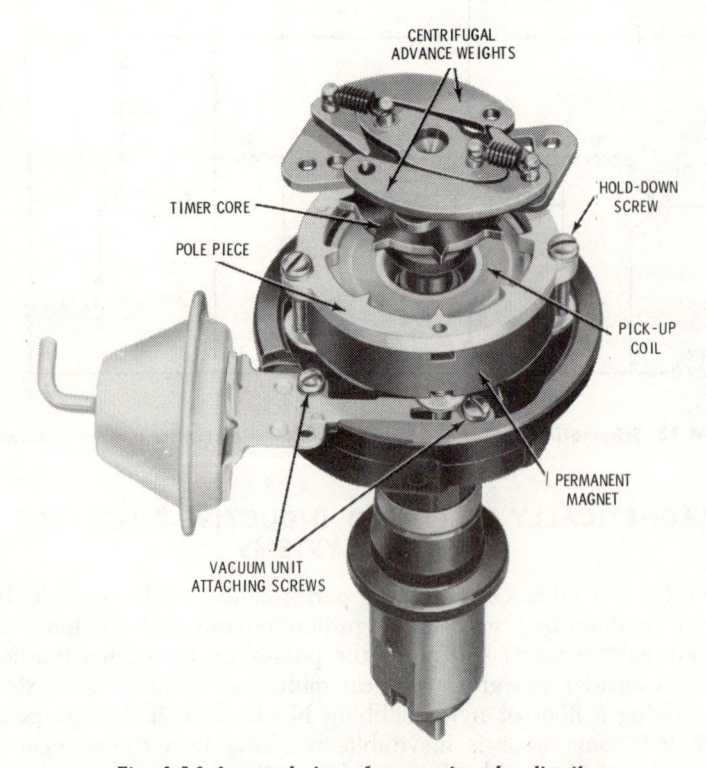

Fig. 4-16. Internal view of magnetic pulse distributor.

a number of vanes and is attached to the distributor drive shaft. A pickup coil surrounds the central timer core and is enclosed within a pole-piece assembly incorporating a circular permanent magnet. The pole-piece assembly has a number of internal teeth corresponding to the engine cylinders.

This assembly provides triggering pulses to an external control unit known as an ignition-pulse amplifier. As the timer core rotates, its vanes pass near the pole-piece teeth, thereby causing the magnetic field about the pole pieces to alternately intensify and fall off. This periodic change in flux strength cuts the conductors of the pickup

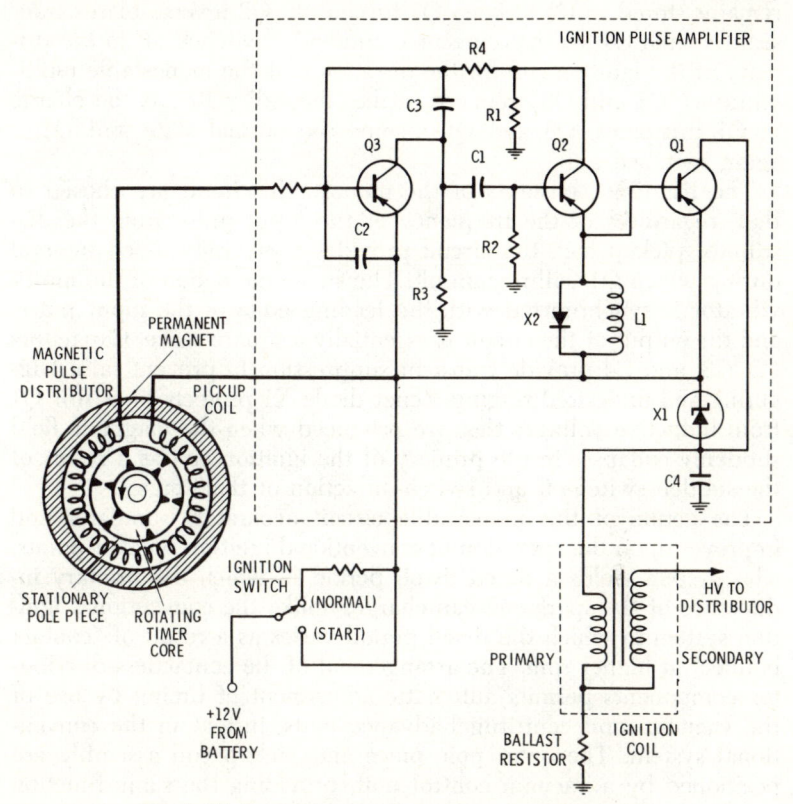

Fig. 4-17. Contactless ignition system by Delco-Remy.

coil, inducing a voltage pulse across the coil each time a timer-core vane passes a pole-piece tooth. Each voltage pulse is conducted to the ignition-pulse amplifier unit, shown schematically in Fig. 4-17.

The pulse-amplifier circuit contains a monostable multivibrator consisting of transistors Q2 and Q3 which controls the state of transistor switch Q1. The circuit is so arranged that, with no input from

49

the pickup coil, transistor Q3 is not conducting but transistor Q2 is biased into conduction by the current through resistor R2. This is the so-called "normal" state. Transistor switch Q1 is forward biased by the emitter-collector current of Q2, allowing current to flow through the ignition switch, the ballast resistor, and the primary of the ignition coil. When an incoming pulse from the pickup coil arrives at its base, transistor Q3 conducts and transistor Q2 is momentarily reverse biased by the pulse coupled to its base by capacitor C1. As Q2 turns off, Q3 receives forward bias through resistors R1 and R4 in series and remains conducting so long as a sufficient charge remains stored in C1 to keep Q2 turned off. Q2 reverse biases transistor switch Q1, so that current is suddenly switched off in the primary of the ignition coil for the duration that the monostable multivibrator (Q2 and Q3) remains in the "flipped" state. As the charge in C1 dissipates, Q3 and Q2 assume the normal state and Q1 is again switched on.

The RC time constants of the monostable circuit are chosen so that, regardless of the frequency of the input pulse from the distributor pickup coil, the circuit provides a precisely timed interval during which Q1 will remain off. The switching action of the multivibrator is synchronized with the leading edge of the input pulse, and the output of the circuit is essentially a square wave. Capacitors C2, C3, and C4 provide transient suppression to prevent false triggering and undesired ringing. Zener diode X1 protects transistor Q1 from inductive voltages that are produced when the magnetic field suddenly collapses in the primary of the ignition coil as a result of the sudden switch-off and switch-on action of that stage.

The action of this monostable circuit accurately simulates and improves upon the operation of conventional ignition breaker points. This system yields a timed dwell period in which the primary inductance of the spark coil can charge, unlike the conventional ignition system in which the dwell period varies as a result of "contact bounce" at higher rpm. The arrangement of the contactless-distributor components permits automatic adjustment of timing by use of the vacuum- and centrifugal-advance units, just as in the conventional system. The outer pole piece and pickup coil assembly are positioned by a vacuum control unit, providing the same function as the vacuum-advance mechanism in the conventional ignition system. The central timer core is rotated about the distributor shaft by the centrifugal-advance weights to provide the same centrifugal-advance function as in the conventional ignition system.

Chrysler Magnetically Triggered Electronic Ignition System

Credit for the first *production* solid-state ignition system as standard equipment in motor vehicles properly goes to the Chrysler Corporation. All of its 1973 vehicles featured electronic ignition as standard equipment (Fig. 4-18), though substantially the same system had been available since 1971 as a special option on some lines. Although there is a measure of similarity between the early Pontiac system (just described) and the Chrysler system, there are important differences. For one, the magnetic pickup unit in the Chrysler distributor is quite unlike the earlier Pontiac unit. In the distributor of the Chrysler electronic ignition system (Fig. 4-19), a small permanent magnet in the pickup unit provides a magnetic field. This magnetic field surrounds a coil that is wound around the pole piece. The field is relatively weak because the air gap between the pole

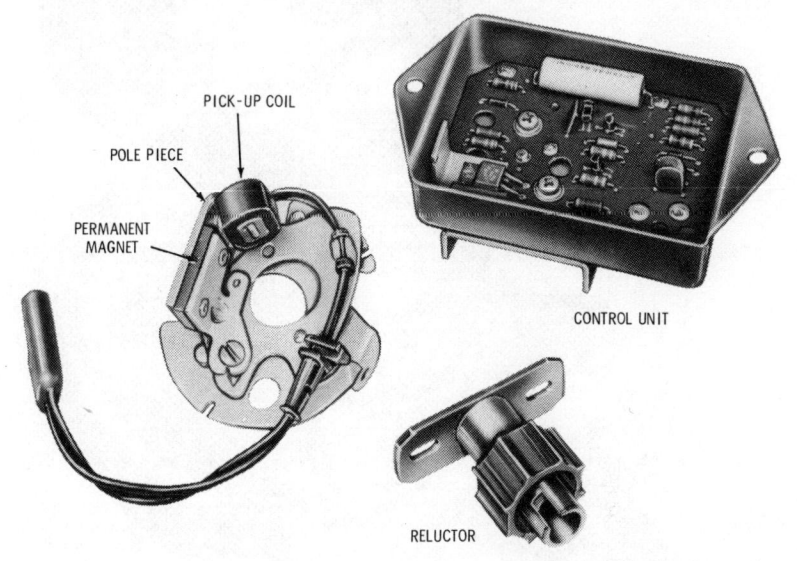

Fig. 4-18. Components of Chrysler electronic ignition system.

piece and the magnet does not provide a good magnetic path between the two (Fig. 4-20).

The pickup unit is mounted close to the rotating *reluctor* which has teeth corresponding to the number of engine cylinder. As a tooth of the reluctor approaches the pickup, it provides a better magnetic path than the air gap and the strength of the magnetic field in the pickup coil is increased. Increasing the field strength at the pickup coil includes a positive voltage at one terminal of the coil. It should be understood that this voltage is induced as a result of the changing and increasing field strength and is not caused by movement of the field or the pickup coil. The positive voltage continues to build until the reluctor tooth is exactly opposite the pole piece as shown in Fig. 4-21.

As soon as the reluctor tooth passes the pole piece, the air gap starts to increase and the field strength begins to decrease (Fig. 4-22). The decreasing field strength through the coil winding induces a negative voltage at the same terminal of the coil winding

Courtesy Chrysler Corporation

Fig. 4-19. Distributor used with Chrysler electronic ignition system.

that was positive with a strong field. Again, the voltage is induced by the change (reduction) in field strength. No voltage is induced in the pickup coil unless the *reluctor is moving*. The rapid increase and decrease of the magnetic field as the rotating reluctor teeth approach and pass the pole piece produces the changing voltage.

The induced voltage is just a tiny electrical signal that is fed into the electronic control unit. The function of the signal voltage induced in the pickup unit is not the same as that of the breaker points in a conventional ignition system which open and interrupt the primary current in the ignition coil. The pickup voltage is a precisely timed command signal. It triggers the electronic circuitry in the control unit which, in turn, controls the interruption of the current flowing through the primary winding of the ignition coil. The function of the control unit is shown in Figs. 4-23 and 4-24.

In the Chrysler electronic ignition system, battery current flows through the primary winding of the ignition coil and then through the control unit shown in Fig. 4-25. The control unit maintains current flow in the primary winding of the ignition coil much the same as the closed contacts do in a conventional Kettering ignition system. The control unit remains on, or activated, and current flows through the primary coil windings as long as a negative voltage from the pickup is *not* applied to it.

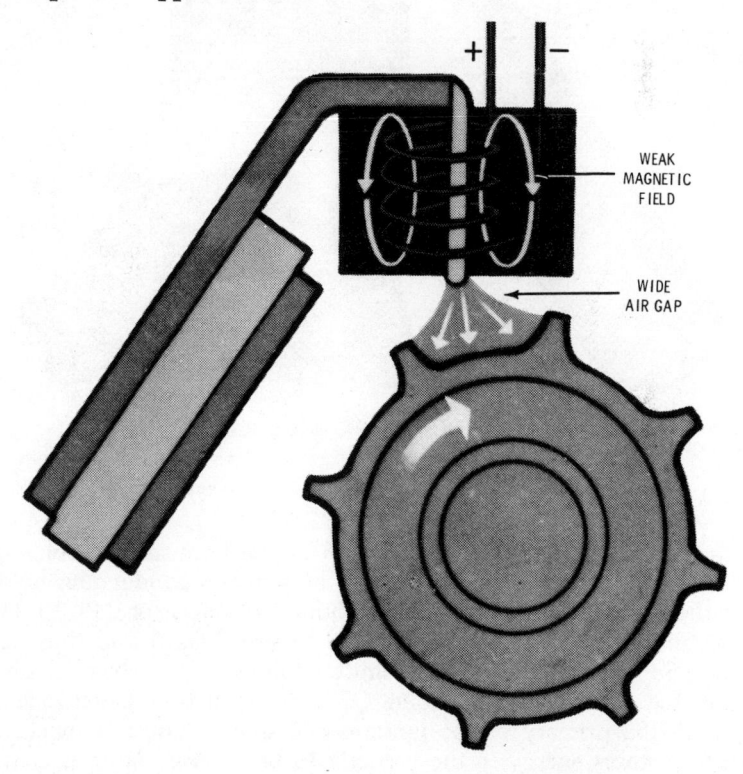

AIR GAP OFFERS RESISTANCE TO FIELD

Courtesy Chrysler Corporation

Fig. 4-20. Weak magnetic field due to wide air gap.

53

When the reluctor passes the pole piece, the pickup voltage becomes negative and deactivates or turns off the control unit circuitry. At this point, current cannot flow through the control unit to ground and, therefore, the current through the ignition-coil primary winding is interrupted. As in all inductive-discharge ignition systems, this interruption of the current flow in the primary circuit of the ignition coil induces enough voltage in the secondary winding to fire a plug.

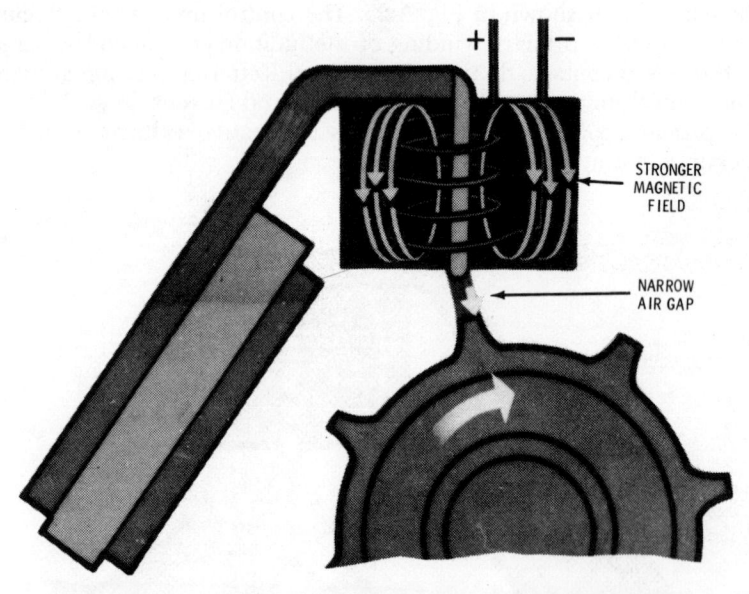

INCREASING FIELD STRENGTH INDUCES POSITIVE VOLTAGE

Courtesy Chrysler Corporation

Fig. 4-21. Stronger magnetic field due to narrow air gap.

Refer to the schematic in Fig. 4-26 to see how the control unit operates. Ordinarily, transistor Q1 is biased into conduction, holding the anode of programmable unijunction transistor (PUT) Q2 at essentially ground potential. Q3 is reverse biased and thus applies a forward-bias current to emitter follower Q4, which in turn biases the power switching stage Q5 into conduction. Current flow through the primary of the ignition coil thus establishes the field flux and stores energy in the inductor to be released later as spark potential.

As a moving reluctor tooth cuts the field of the pickup coil magnet, the shift in flux generates a current flow in the circuit consisting of L1, D1, and D2. This current flow supplies a negative-polarity

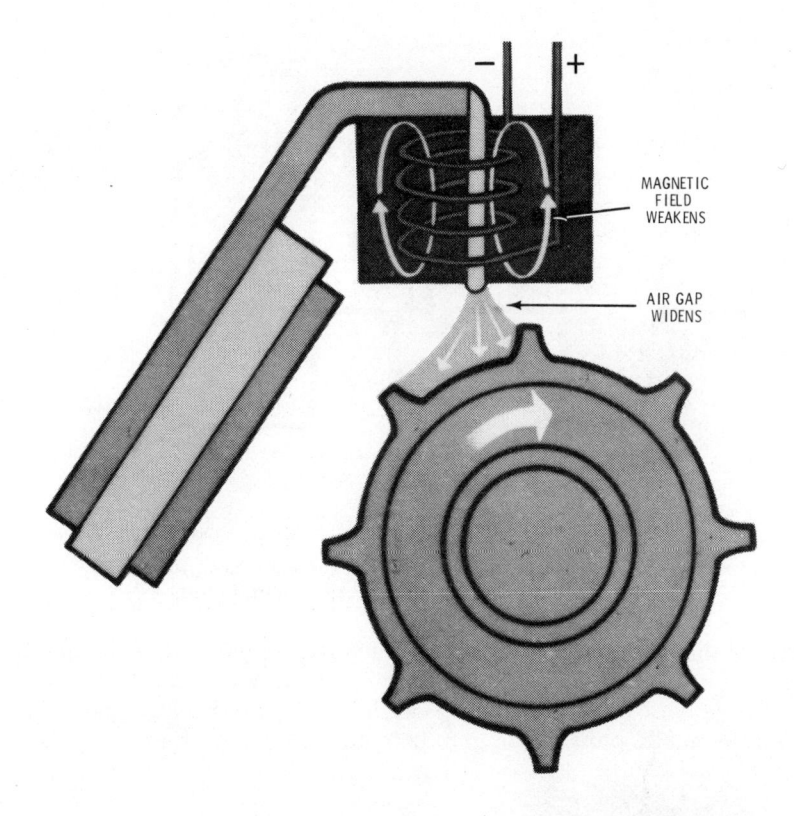

Courtesy Chrysler Corporation

Fig. 4-22. Weaker field produces negative voltage as tooth passes the pole piece.

pulse to the base of Q1. The amplitude of this pulse is sufficient to momentarily drive Q1 out of conduction, thereby applying a positive pulse to the anode of PUT Q2. The gate of the PUT is biased

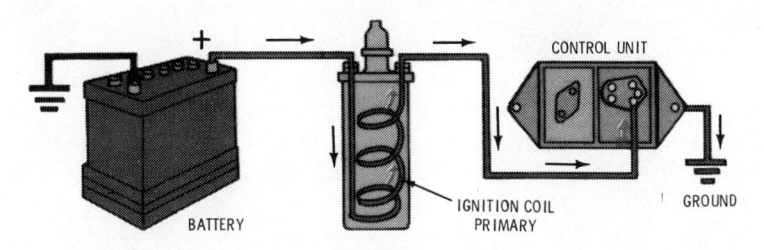

Courtesy Chrysler Corporation

Fig. 4-23. Ignition-coil primary current flows through control unit.

55

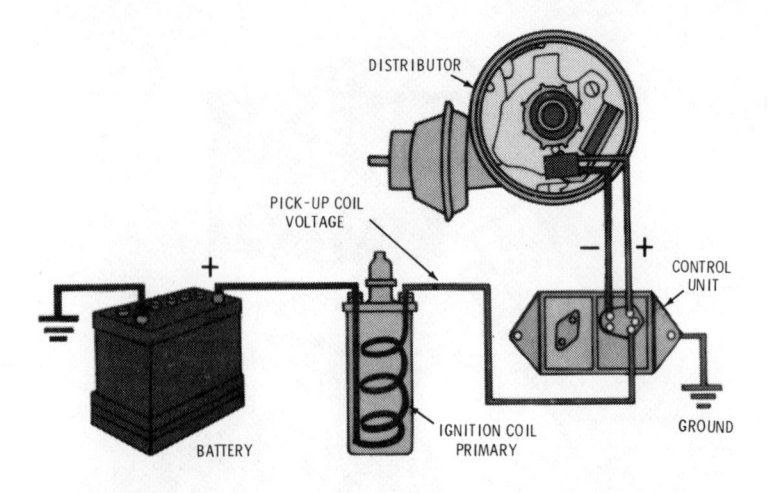

Fig. 4-24. Negative voltage from pickup coil causes control unit to interrupt current through primary of ignition coil.

by the divider network R4 and R5, such that the arrival of the positive pulse from Q1 causes the gate of Q2 to momentarily appear negative with respect to the anode. Instantly, the PUT triggers on, and the anode pulls to the gate potential, coupling a positive pulse

Fig. 4-25. Chrysler electronic ignition control unit.

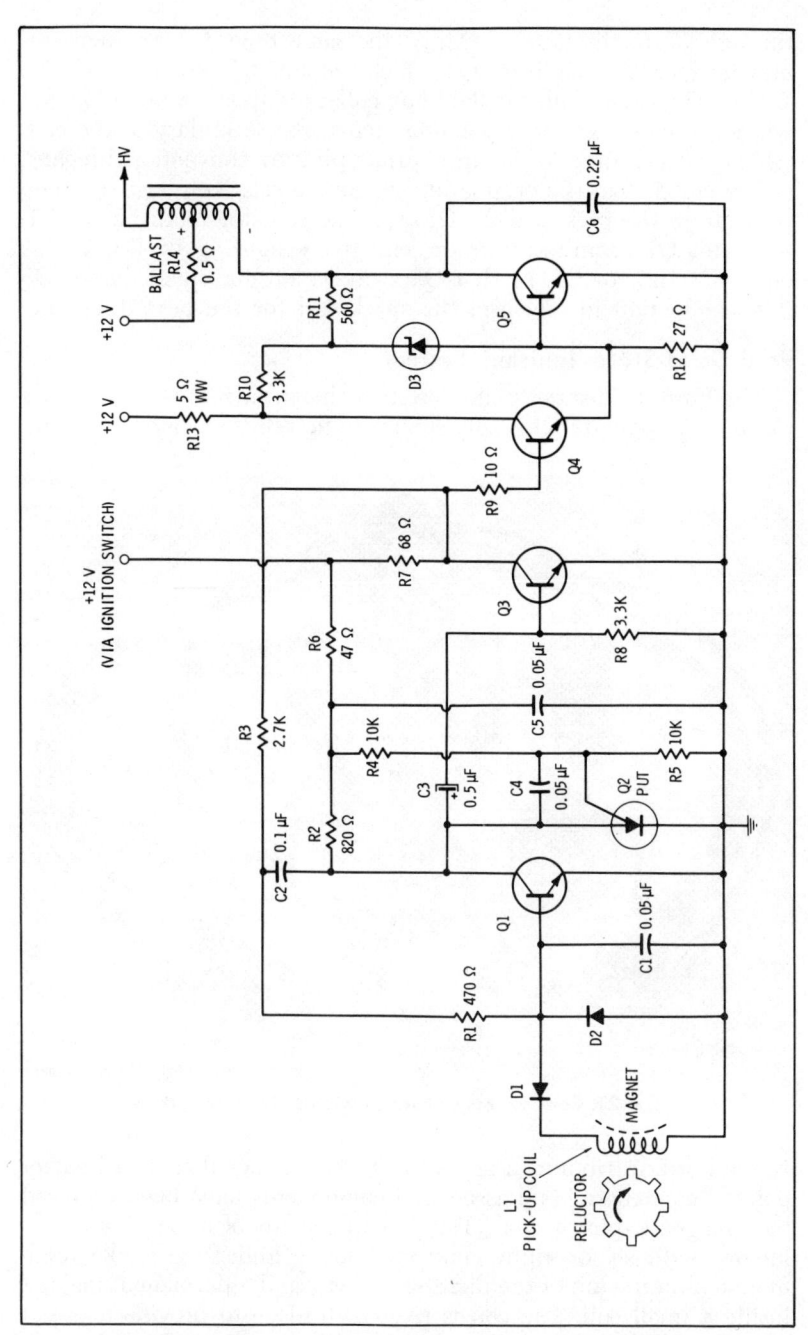

Fig. 4-26. Schematic of Chrysler electronic ignition system.

through C3 to the base of Q3. At the same time, Q3 conducts, removing forward bias from Q4 which immediately switches off Q5. When Q5 switches off, the field flux collapses about the ignition-coil windings, inducing spark potential across the secondary of the coil, which is distributed to the appropriate plug by conventional means.

Remember that the originating signal is a relatively fast rise-time pulse from the pickup coil (Q1). As the reluctor tooth passes L1, transistor Q1 again switches on, effectively shorting PUT Q2. This cuts off transistor Q3, so that Q4 switches on and again biases Q5 into conduction to recharge the spark coil for the next discharge.

Ford Solid-State Ignition System

The Ford solid-state ignition system shown in Fig. 4-27 utilizes a principle of operation broadly similar to that of the Chrysler system.

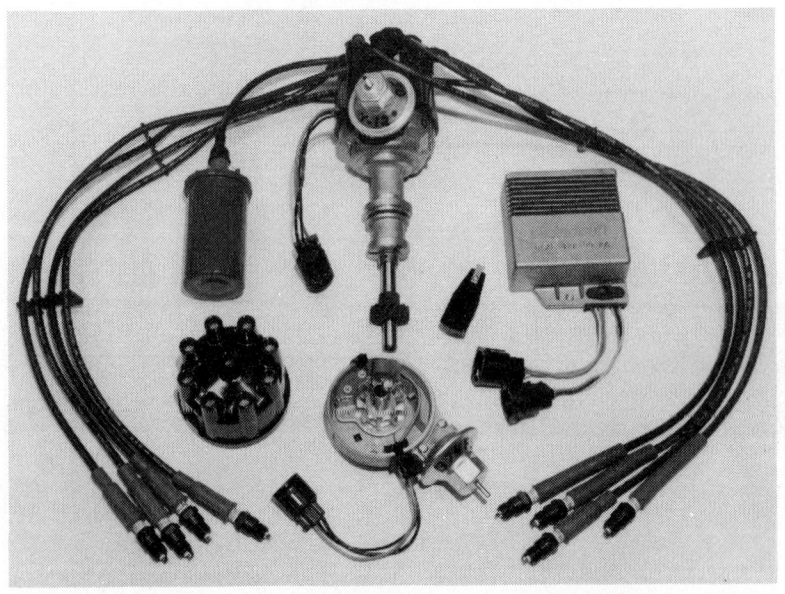

Courtesy Ford Motor Company

Fig. 4-27. Components of Ford solid-state ignition system.

In the Ford distributor (Fig. 4-28), the conventional contact breaker points, "condenser," and associated components have been replaced by a magnetic-pickup unit. The pickup consists of an armature (reluctor) with six or eight (one for each cylinder) gear-like teeth mounted on the top of the distributor shaft and a permanent magnet inside a small coil. The coil is riveted in place to provide a preset air gap with the reluctor. The distributor base, cap, rotor, and spark-

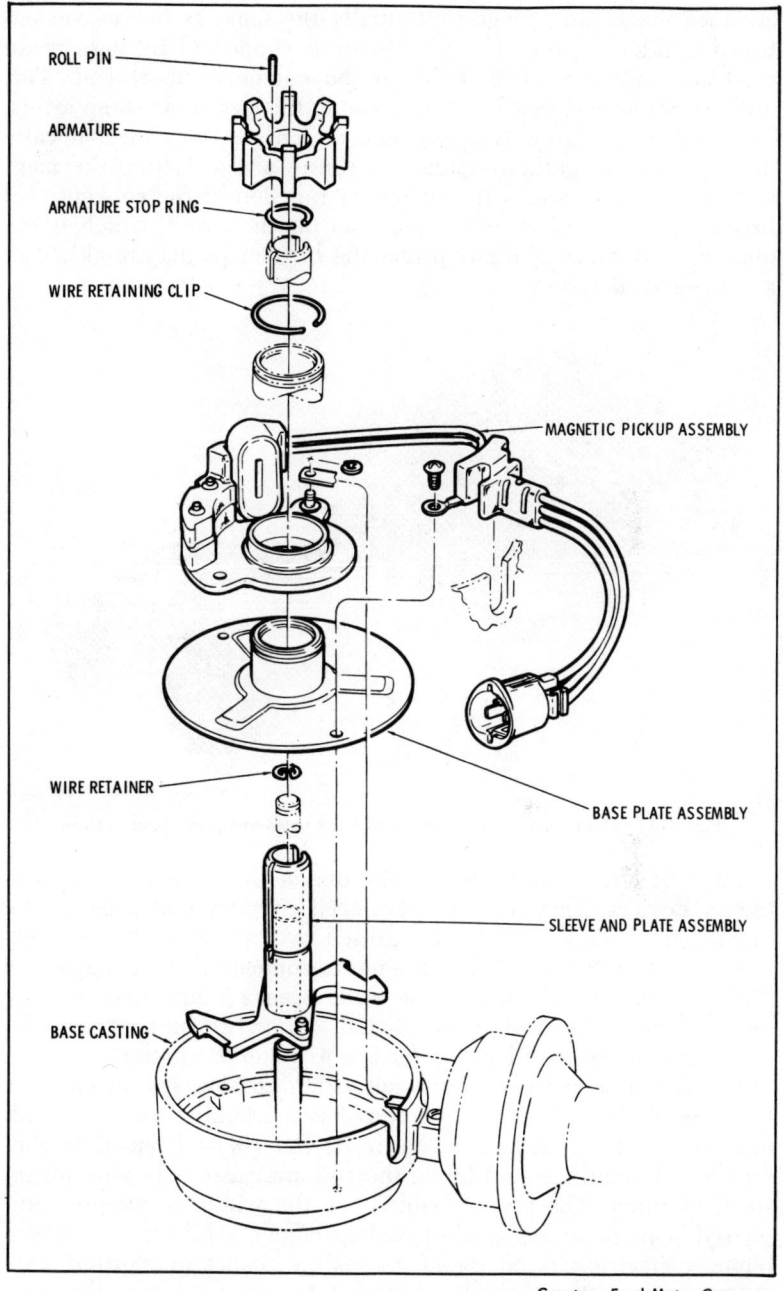

Fig. 4-28. Ford magnetic-pickup distributor.

advance mechanism remain essentially the same as in the conventional ignition system. The distributor is connected by wire to an electronics module (Fig. 4-29) in the engine compartment. This module houses a circuit board containing the electronic components. The module performs two functions which are done mechanically in conventional ignition systems. It senses a signal from the magnetic pickup to perform the switching function of the conventional breaker points, and it senses and controls the dwell, which is the number of degrees of rotation that the breaker points are closed in a conventional system.

Courtesy Ford Motor Company

Fig. 4-29. Electronics module used in Ford solid-state ignition system.

Some of the components within the module are used as protective devices. They are part of special circuitry that protects the module from damage or failure caused by reverse polarity or high voltage. These two conditions are not uncommon during emergency starting. For example, the wrong leads from a jumper battery may be accidentally touched to the car battery, or two batteries may be connected in series (24 volts) for use as a jumper battery.

The coil used with the Ford solid-state ignition system employs the same design principles as a conventional ignition coil, but with new construction. Major differences are the use of a plastic bobbin for the coil windings and filling the coil after assembly with oil instead of pitch. The plastic bobbin for the windings provides improved manufacturing control and assembly, while the oil has a higher dielectric efficiency that lessens the chance of shorting.

During operation of the Ford solid-state ignition system, the magnetic pickup inside the distributor sends a simple alternating-current signal to the module. The current waveform swings from positive (+) to negative (−) each time one of the gear teeth on the armature passes the permanent magnet in the coil. When a gear tooth is exactly opposite the coil, the current waveform is at a crossover (zero). The electronic module is designed to sense this position and cut off the primary current to the spark coil, causing it to fire. This is the same function that occurs when the breaker points open in a conventional ignition system. Timing circuitry in the module also senses when the coil has fired and then restores the primary current to the coil. This is the same function performed by the closing of the contact points in a conventional system. The high voltage from the secondary of the ignition coil is directed to the spark plugs through the distributor, as in the conventional system. Automatic advance or retard of the spark is controlled the same way as in a conventional system, except that, instead of moving the breaker-point plate, the plate mounting the pickup coil is moved.

Fig. 4-30 is a schematic of the Ford solid-state ignition system. As the input signal waveform crosses through zero in a positive-going direction, transistor Q1 is turned on. With a current path provided by Q1, capacitor C1 begins to discharge through D7. As this is happening, the base of Q2 is reverse biased and Q2 is turned off. The time required for this to take place is called *spark time*, which is a function of the R4C1 time constant and the frequency of the input signal from the magnetic pickup.

As Q2 is turned off, its collector rises to approximately 1.4 volts which is fed back to the base of Q6, turning it on and providing another path for the C1 discharge current should Q1 turn off due to noise in the signal. Even though Q1 is always on for more than 50% of the time, noise on the incoming-signal lines could possibly turn it off and shorten the burn time of the spark during its critical period of operation. Hence, a latch action is provided for Q2 to ensure sufficient spark time.

As C1 recharges through R3, the base of Q2 becomes positive and Q2 is turned on again. This is the *dwell computation* which is a function of the R3C1 time constant and the frequency of the input signal. The frequency range of 3 to 400 Hz, which corresponds to an engine speed of 45 to 600 rpm, results in a wide variation in dwell time. However, at engine speeds of 600 to 6000 rpm, the dwell time corresponds very closely to that of the conventional ignition system.

Now that spark time and dwell functions are present in the signal, we can proceed to the next stage. As transistor Q2 is turned off, Q3 is turned on because its base is driven positive by the square-wave output from Q2. The Q3 stage provides signal inversion and gain needed for the Q4 Darlington stage. The potential at the base of Q4 drops as Q3 turns on, turning transistor Q4 off. The output of

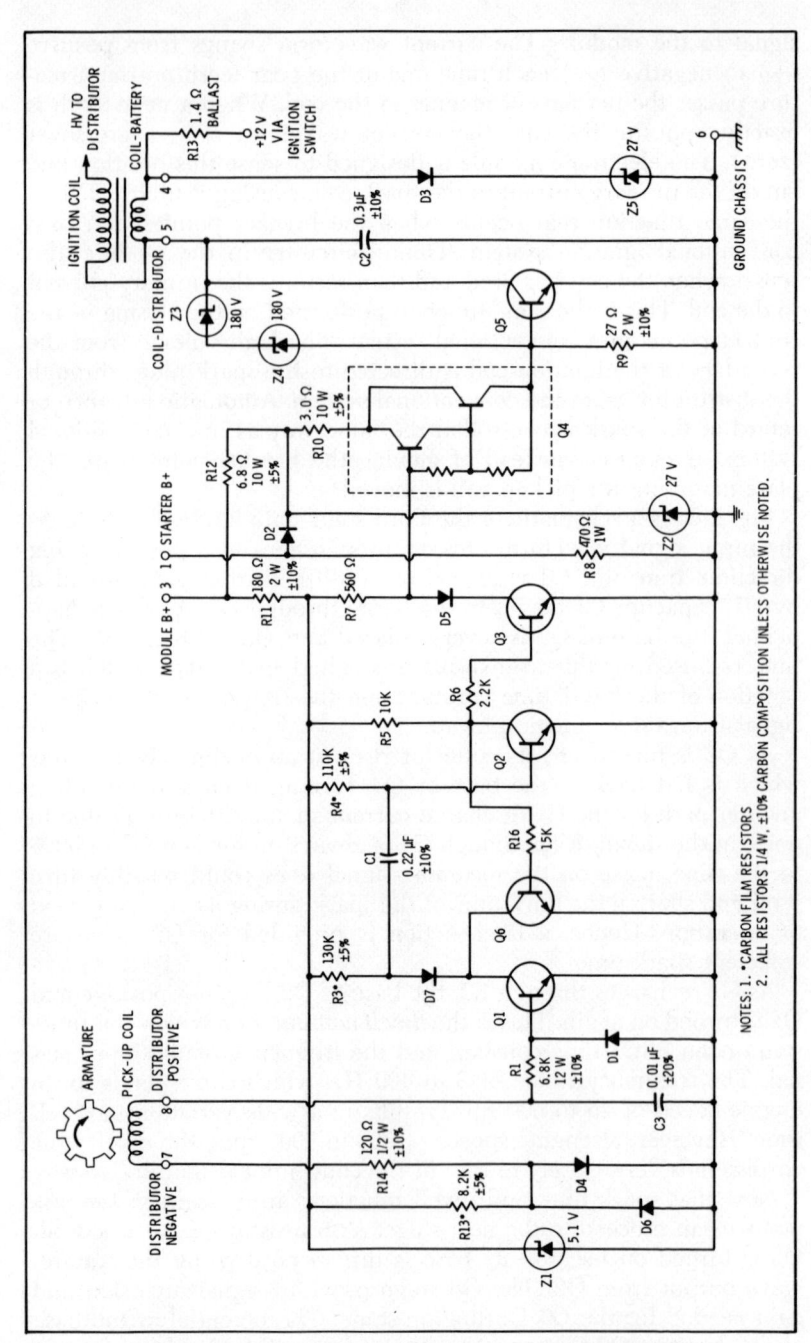

Fig. 4-30. Schematic of Ford solid-state ignition system.

Q4 is then coupled to output stage Q5, where power gain is achieved. The collector of Q5 rises to approximately 360 volts, depending upon capacitor C2 and ignition-coil parameters. As the primary current collapses to zero, the rate of voltage rise (dv/dt) of the ignition-coil primary inductance is the source of the high voltage at the collector of Q5. This voltage never exceeds 380 volts due to the clamping action of zener diodes Z3 and Z4 which protect Q5.

The threshold, or firing point, at which Q1 will turn on is related to the match of the base-emitter voltage of Q1 and the forward voltage drop of D4. The time elapsed from Q1 turn-on to spark plug firing ranges from approximately 8 to 20 microseconds. During engine cranking, resistor R12 is bypassed by applying the A line (battery voltage) to terminal 1, thus increasing the current through Q4 and the drive to output stage Q5. The 1.2-ohm ballast resistor (R13) is also removed from its series path with the primary of the coil during cranking. These two conditions combine to provide increased energy at the spark plug during engine cranking and to help give easy starts.

General Motors/Delco-Remy Unitized Ignition

Delco-Remy's new unitized ignition system combines the ignition coil, the secondary wiring harness, and the magnetic pulse distributor into one compact unit. The new system has been offered as an option on certain Pontiac 455-cu. in. engines and it will see further applications on other General Motors cars, according to company sources.

The new unitized ignition system supposedly offers a significant advance in reliability and performance over conventionally used separate components. By incorporating solid-state integrated-circuit technology, the ignition coil and the secondary wiring harness are made an integral part of the distributor cap. They form a completely sealed and waterproof assembly. With the unitized system, there is 1 unit to install and 9 electrical connections to complete; whereas, with conventional ignition, there are 12 components and 21 connections. The unit is approximately the same size as a conventional ignition distributor. The separate coil and its mounting space are eliminated, along with the wiring harness between distributor and coil.

The new system completely eliminates the cam, rubbing block, points, and condenser which are necessary with a conventional distributor. Instead, a magnetic pickup triggers a hybrid electronics module to switch power to the ignition coil, making this the first production inductive-discharge electronic ignition system to employ a custom IC (integrated circuit). More voltage is available for ignition during starting, spark plugs last longer, and performance deterioration is eliminated with the result that periodic distributor maintenance is virtually eliminated, according to GM spokesmen.

Fig. 4-31 shows the components of the unitized system. Unlike the Chrysler and Ford magnetic pickups, the Delco design employs a pickup pole piece having salient poles, matching the number of teeth on the timer-core reluctor. The pickup design is reminiscent

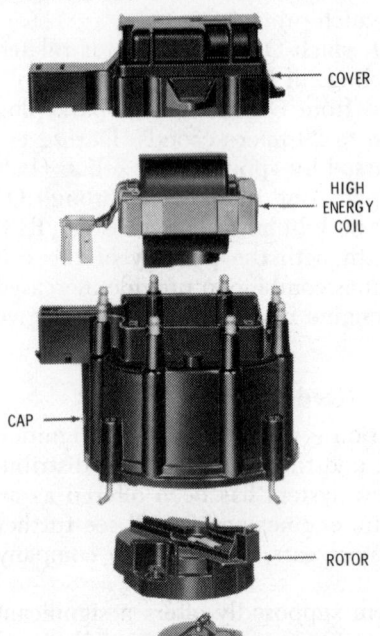

Fig. 4-31. Components of GM/Delco-Remy unitized electronic ignition system.

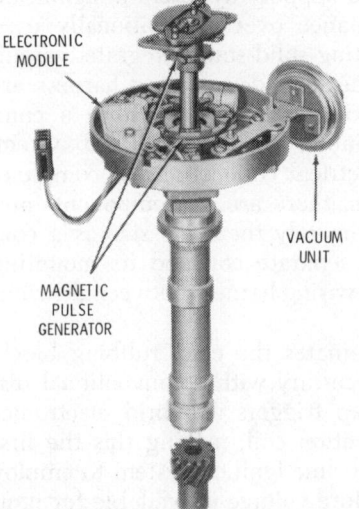

Labels: COVER, HIGH ENERGY COIL, CAP, ROTOR, ELECTRONIC MODULE, VACUUM UNIT, MAGNETIC PULSE GENERATOR

Courtesy General Motors Corporation

of the 1963 Delco system described earlier. Use of multi-pole coupling improves the output voltage of the pickup at low cranking speeds.

The electronics of the Delco unitized system are modularized in a hybrid IC package (Fig. 4-32) that mounts below the distributor shell. Connections between the IC and the primary of the coil are made through blade-type, quick-disconnect fasteners. As shown in Fig. 4-31, the coil is integrated with the distributor cap. The coil secondary circuit and spark plug wires terminate in a header fitted with contacts that protrude through the distributor cap shell into the interior of the distributor. There, the rotor atop the timer-core shaft distributes spark potential to the correct plug in the conventional manner.

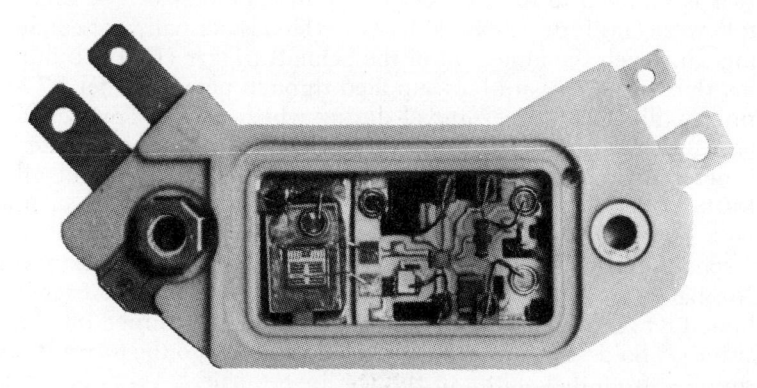

Fig. 4-32. Electronics module used in GM/Delco-Remy unitized ignition system.

Circuit Description

A schematic of the Delco-Remy unitized ignition system is shown in Fig. 4-33. The circuitry within the dashed lines is part of the monolithic IC (integrated circuit) housed within the electronics module. Other circuit components are contained within the module but are separately supported.

Operation of the magnetic pickup unit is essentially identical to that described earlier in this chapter for the pioneering efforts in magnetically triggered ignition systems by Delco-Remy in 1965. However, the electronics are substantially different. Basically, the HEI ignition module contains: a biasing circuit (Q1) to assure adequate energy storage in the pickup coil regardless of engine rpm; a Schmitt trigger (Q2 and Q3) to convert the sinusoidal voltage generated by the magnetic pickup unit into control-current pulses; an emitter follower (Q4) to isolate the Schmitt trigger from the

65

stages that follow; a high-voltage Darlington switch (Q5 and Q6), which controls the flow of primary current through the ignition coil; and switching control transistors (Q7, Q8, and Q9), which shape the switching pulses for optimum operation.

The Darlington connection of Q5 and Q6 creates a very high gain switch capable of faster response time than a single transistor. This is because the current gain of Q5 is multiplied by the current gain of Q6. The Darlington switch is capable of very fast switching speed, which results in a lower operating temperature for the device. It is this feature that allows the switching circuit to be packaged within the confines of the electronics module.

The Darlington switch is conducting during periods when a positive half-cycle of signal is supplied by the magnetic pickup. Forward bias is supplied to the base of Q5 from the emitter of Q4. Emitter-follower Q4 is forward biased by Q3, which is normally not conducting. However, the input half of the Schmitt trigger (Q2) is conducting due to the forward bias supplied through potentiometer R7 and resistor R8. This is the interval during which current flows to store energy in the primary of the ignition coil. The period of the interval is determined by the time required for the timer core of the distributor to traverse the distance from one point of alignment to another with respect to the fixed teeth of the pickup unit. These points correspond to peak flux and, thus, peak current through the pickup coil. Normally, diode D1 is not conducting and the Schmitt trigger is biased into conduction with Q2 turned on and Q3 turned off. Transistor Q8 holds the collector of Q3 turned off due to the forward bias supplied from the emitter of the Darlington pair.

Let us assume that the pickup timer core approaches a new alignment with the fixed teeth, inducing a peak signal of negative polarity in the pickup coil. Instantly, diode D1 conducts, causing transistor Q2 to turn off. Regenerative coupling through resistors R9 and R10 immediately turns transistor Q3 on, causing Q4 to be reverse biased and turning off the Darlington switch (Q5 and Q6). Instantly, the primary current through the ignition coil is interrupted and a high voltage is induced in the secondary of the coil. Since the full supply voltage is essentially available at the collector of Q4 during this interval, transistor Q7 is turned on by the forward bias developed by the divider network consisting of R15 and R16. Q7 holds the base voltage of Q5 below the turn-on level of the Darlington switch (nominally 1.3 volts), thus preventing premature termination of the spark time by transients. As the timer core passes the alignment point and the signal polarity reverses, diode D1 ceases to conduct and the state of the Schmitt trigger reverses. This switches the Darlington stages into conduction again to charge the ignition coil in readiness for the next spark-plug firing.

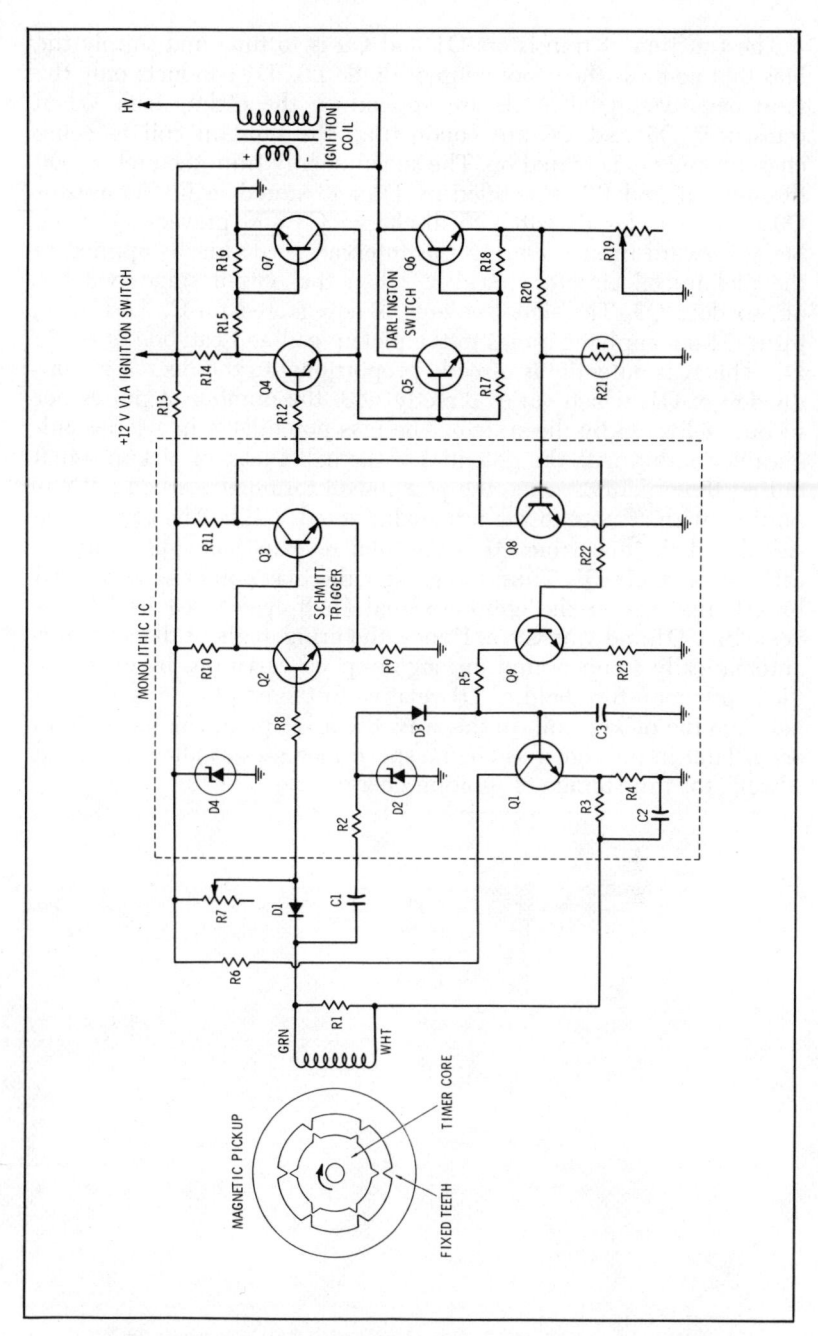

Fig. 4-33. Schematic of GM/Delco-Remy unitized ignition system.

The function of transistors Q1 and Q9 is to time and supply the bias that controls the conduction of diode D1. D1 conducts only the most negative signal levels are applied by the pickup coil. When transistors Q5 and Q6 are conducting, the ignition coil is being charged and Q9 is biased on. The signal energy from the pickup coil through C1 and R2 is rectified by D3 and stored in C3. Transistor Q9 acts as a closed switch to discharge C3 and prevent Q1 from being forward biased. During this interval, no dc bias is applied to the pickup-coil circuit. However, when the output stage switches off, so does Q9. The signal voltage is now stored on C3 and, thus, turns Q1 on, applying a bias to the pickup coil and cathode of diode D1. This bias potential is directly proportional to the degree of conduction of Q1, which varies directly with the number of pulses per second delivered by the system. The bias potential is in a series-aiding relationship with the potential of the half-cycles of pickup signal during those periods when the pickup-coil terminal marked GRN is positive with respect to the terminal marked WHT. The sum of the signal and the bias from Q1 is the total reverse bias applied to the cathode of diode D1. Thus, the greater the bias potential developed by Q1, the smaller the ignition signal (half-cycle) required to reverse bias D1 and vice versa. Hence, the firing angle of the system is automatically compensated for engine-speed variations by changing the conduction threshold of D1 relative to the amplitude of the signal from the pickup coil. In this way, the firing point can be made to occur later in the signal as engine-speed increases to allow sufficient "dwell" for full storage of ignition energy.

CHAPTER 5

Capacitive-Discharge Ignition Systems

Not unlike a host of other useful innovations, a *capacitive-discharge* ignition had its birth in the terrible crucible of war. During World War II, aircraft designers for the Nazi regime recognized that the skies belonged to the combatant whose aircraft were first off the ground. Immediately reliable ignition without engine warmup was their goal, and they achieved it through a capacitive-discharge ignition system. It had long been known that energy could be stored in the *electric field* of a capacitor, just as it could be stored in the magnetic field of an ignition coil. From this starting point, the German designers evolved a crude system in which a capacitor was charged with a high dc voltage, then discharged through a gas tube into the ignition coil to provide a hot spark to a cold aircraft engine. The means used by these early designers were clumsy—a rotary converter and fragile tube circuitry, which were badly suited to the harsh environment of a fighter plane. Fortunately, component problems prevented this ignition system design from becoming standard on Axis aircraft; otherwise, the course of history might have been changed enormously.

In the late 1950s, a Tung-Sol division, known as Motion Incorporated, briefly revived and experimented with a CD ignition system concept for automobiles. But they also abandoned the effort because available tubes and power supply components lacked staying power. A CD ignition system typical of early designs employing an electron tube for switching is shown in Fig. 5-1.

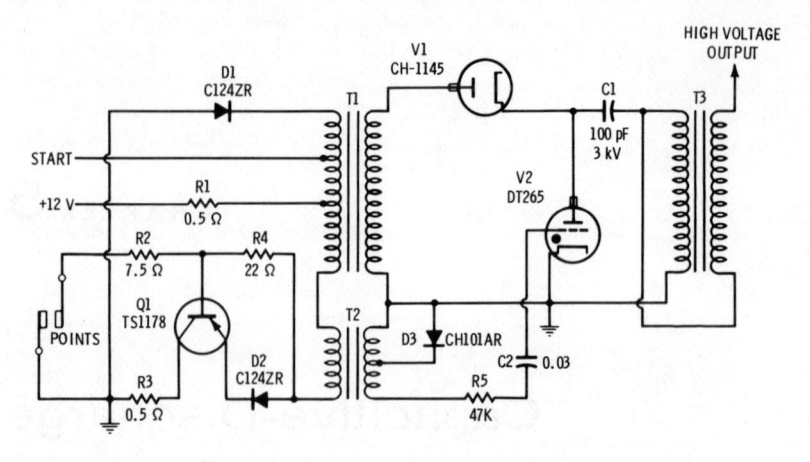

Fig. 5-1. Early gas-tube CD ignition system.

HOW CAPACITIVE-DISCHARGE IGNITION SYSTEMS OPERATE

When current flows into a capacitor, a dielectric flux is established between two parallel plates separated by an insulator. In effect, electrons have piled up on one plate and their negative charge has repelled a like number of electrons on the opposite plate. Thus, energy is said to be stored in the capacitor. Unlike, an inductor, this energy remains stored in the capacitor when input current flow ceases. Connecting a conductor across the capacitor provides a plate-to-plate path by which the charged capacitor can regain electron balance (become discharged). A moderate-size capacitor can store an appreciable electron charge; therefore, a large-scale discharge current is possible if the conductor shunting the capacitor plates has low inductance and resistance. The equation for determining energy stored in a capacitor is:

$$e = \frac{CE^2}{2}$$

where,
    e is energy in joules,
    C is capacitance in farads,
    E is voltage applied across the capacitor in volts.

Let us assume that the charged capacitor is placed across the primary winding of an ignition coil. Current flows through the primary, and a rapidly propagating magnetic field spreads outward, cutting the secondary turns and inducing a high-voltage pulse. Thus, the ignition coil acts as a pulse transformer, rather than as a charging inductor. The discharge pulse from the capacitor is stepped up to spark-voltage level and energy delivery is immediate. To assure rapid recharging of the capacitor, it is necessary to electrically disconnect it from the coil and charge it with a higher potential than the usual 12-volt supply found in cars.

All present capacitive-discharge ignition systems utilize a silicon controlled rectifier as the switching element. Thus, the main distinction between available CD ignition systems is the type of power supply used to charge the storage capacitor between plug firings. Automotive CD systems almost universally employ a transistorized high-frequency inverter. Smaller engines use a rotary magnetic generator. Typically, the energy-storage capacitor of a CD system has a value of 1 to 2.5 microfarads and the supply voltage is in the range of 350 to 400 volts. The energy delivered into the primary of a practical ignition coil from such a combination is on the order of 60 to 150 millijoules.

INVERTER-TYPE CD IGNITION SYSTEMS

The block diagram of an inverter-type capacitive-discharge system is shown in Fig. 5-2. As shown in the block diagram, the system consists of a dc-to-ac inverter to change the battery voltage to a higher voltage, a bridge rectifier, a storage element (capacitor), a

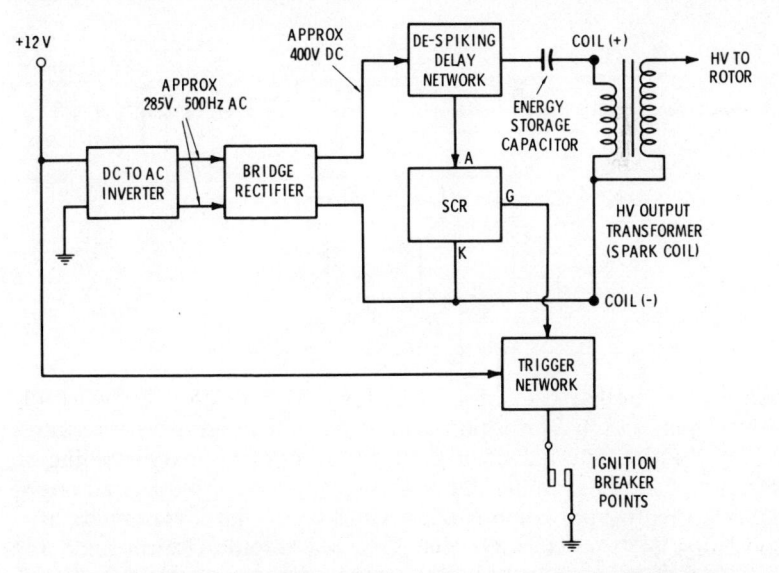

Fig. 5-2. Block diagram of a typical inverter-type capacitor-discharge ignition system.

switching element (SCR), and a high-voltage output transformer (the spark coil) to transform the stored dc voltage to a level that will fire the spark plug.

The objective of the inverter is to establish a supply source which, with rectification, can quickly build up a charge in an energy-storage capacitor in the brief interval between discharges. Only by starting from a fairly high voltage is it possible to store adequate energy in a capacitor of reasonable size in a short time span. Remember, a capacitor charges exponentially, starting from zero volts and rising to the level of the supply voltage in a given period of time. How quickly the capacitor reaches peak charge is determined by the capacitance and the effective series resistance and inductance limiting current flow into the capacitor. Since at 3000 rpm an eight-cylinder engine requires 200 ignition pulses per second, it is obvious that a period of only *five milliseconds* is available between plug firings to recharge the capacitor.

Inverter Circuit Operation

The time constraint is easily met by transistor inverter circuits, of which the circuit illustrated in Fig. 5-3 is typical. The dc-to-ac in-

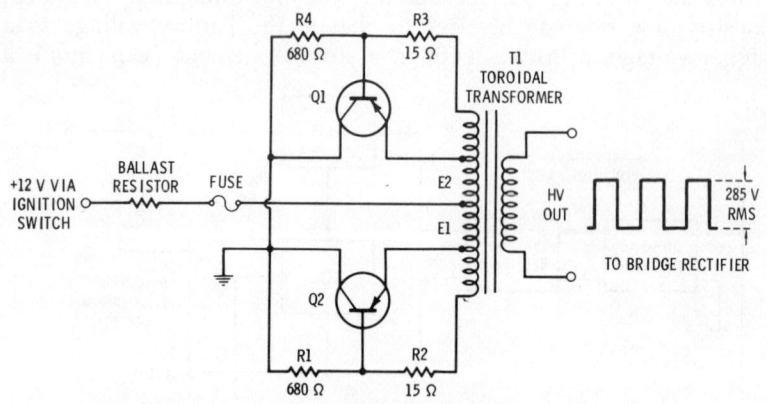

Fig. 5-3. Typical inverter circuit used in CD ignition system.

verter is a familiar transistor oscillator circuit employing two transistors operating in a push-pull arrangement. These transistors switch current to alternate halves of the center-tapped primary winding of a transformer. The applied battery voltage is converted from a nominal 12 volts to approximately 285 volts ac by the inverter circuitry consisting of transistors Q1 and Q2, and toroidal transformer T1. The complete operation of the inverter circuit is explained in the following paragraphs.

The battery voltage applied to transformer T1 causes currents through resistors R1, R2, R3, and R4. Since it is not possible for these two paths to be *exactly* equal in resistance, one-half of the primary winding of T1 will have a somewhat higher current flow. Assuming that the current through the upper half of the primary winding is slightly higher than the current through the lower half, the voltages developed in the two feedback windings (the ends connected to R3 and R2) tend to turn Q2 on and Q1 off. The increased conduction of Q2 causes additional current to flow through the lower half of the transformer primary winding. The increase in current induces voltages in the feedback windings which further drives Q2 into conduction and Q1 into cutoff, simultaneously transferring energy to the secondary of T1. When the current through the lower half of the primary winding of T1 reaches a point where it can no longer increase due to the resistance of the primary circuit and saturation of the transformer core, the signal applied to the transistor from the feedback winding drops to zero, thereby turning Q2 off. The current in this portion of the primary winding drops immediately, causing a collapse of the field about windings of T1.

This collapse in field flux, cutting across all the windings in the transformer, develops voltages in the transformer windings that are opposite in polarity to the voltages developed by the original expanding field. This new voltage now drives Q2 into cutoff and drives Q1 into conduction. The collapsing field simultaneously delivers power to the secondary winding which is connected to a bridge rectifier circuit. This action continues until saturation is again reached, and Q1 ceases to conduct and Q2 is regeneratively turned on through the feedback winding.

This action occurs at a frequency of approximately 500 Hz—the frequency being determined by the saturation characteristics of transformer T1. The 500-Hz square-wave current through the primary of T1 causes a voltage which is approximately 40 times the applied primary potential to appear across the secondary winding of T1.

The high-voltage output of the inverter is applied to diode bridge X1 thru X4 as shown in Fig. 5-4. The rectified dc output from the diode bridge circuit is applied to capacitor C3, charging it to approximately 400 volts through the primary of the spark coil. This action takes place as soon as power is applied, that is, when the ignition switch is closed.

As the first piston comes up on the compression stroke and reaches the position where the spark plug should be fired, the points open and the trigger network supplies a pulse to the gate of the SCR. As the SCR turns on, two things happen simultaneously. First, the SCR short-circuits the output of the diode bridge, with the effect that the

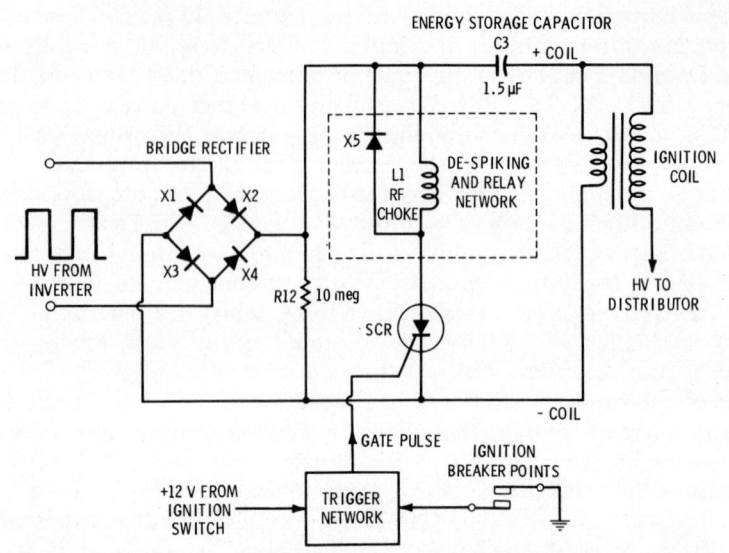

Fig. 5-4. Rectification, energy-storage, and switching circuitry used with CD ignition system.

short is reflected to the primary of T1, removing feedback drive from transistors Q1 and Q2 and instantaneously stopping inverter oscillation. Second, the SCR switches the positively charged plate of C3 to the negative terminal of the spark-coil primary. This forms a closed circuit consisting of the capacitor, the SCR, and the coil primary. The energy stored in the capacitor is now "dumped" into the primary of the spark coil. The primary voltage of the coil rises from zero to 400 volts in two microseconds as shown in Fig. 5-5. Diode X5 and choke L1 limit the energy discharge through the SCR to safe levels.

When the SCR is turned on, a resonant circuit is formed between the spark-coil primary winding inductance and capacitor C3. The flywheel effect of this circuit restores unused energy to the capaci-

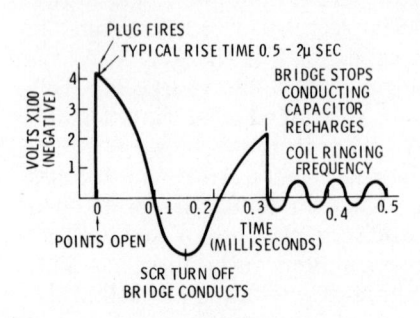

Fig. 5-5. Ignition-coil primary voltage waveform in CD ignition system.

74

tor-discharge current which flows through the SCR and coil primary, creating a magnetic field in the coil. This current continues to flow in the circuit until the capacitor is charged in the reverse direction to approximately 300 volts. At this point, the current attempts to reverse through the SCR, causing the SCR to switch to its off condition. The reverse voltage now causes the diode bridge to conduct (all diodes simultaneously in the conduction mode), thereby draining the reverse charge accumulated in C3. When the current supplied by the ignition coil inductance again drops to zero, the load is removed from transformer T1 and inverter operation resumes, since feedback is now available to start the transistors oscillating.

Performance Characteristics of Inverter-Type CD Systems

The inverter circuit and bridge rectifier will deliver full energy to capacitor C3 at engine speeds exceeding 8000 rpm or well in excess of maximum speed for most engines. Between spark pulses, the inverter has plenty of time to recharge the capacitor. Since the total system current drain is low, the CD ignition system normally operates with the ballast dropping resistor used with most conventional ignition systems. In contrast to a contact-triggered transistor system, this eliminates the need for rewiring the vehicle and/or adding an ignition relay. One of the principal advantages of the CD ignition system is its ability to provide a peak-pulse starting spark to the engine—especially to a cold engine in sub-zero weather. The CD system depends upon dumping a stored energy quantity rather than charging an inductor from the very low 12 volts available from the conventional automotive electrical system. Therefore, it is easy to understand that a substantial pulse output can be obtained even when the starting motor is robbing battery-terminal potential and decreasing it by as much as six volts as the starter attempts to crank a stiff, cold engine.

Another important characteristic of the capacitive-discharge system stems from its use of a silicon controlled rectifier. Remember that an SCR will conduct when its gate is first triggered by an input pulse and remains on so long as holding current is flowing through it. Hence, the CD ignition system is self-completing once an initial input pulse has triggered the SCR into conduction. The circuit will dump its stored capacitor charge through the ignition coil and will continue to drain it "dry" until the SCR is turned off by the reversal of the pulse current in the primary of the ignition coil. This means that dirty, worn, or contaminated breaker points do not degrade the spark energy available to the spark plug. This significant feature contrasts with the all-important need for clean, low-resistance point surfaces in conventional and contact-triggered transistor ignition systems. The SCR gives a clean, definite switching action of the type which Kettering sought long before the invention of this device. This "single-trigger" characteristic of the CD system allows good ignition performance with breaker points that could not be used in either a conventional or contact-triggered transistor ignition system, because of high contact resistance or general degradation.

The short (microsecond) rise time of the CD ignition system will continue to fire fouled plugs long after the transistor and conventional systems (80- to 200-microsecond rise time) have given up. This means a better chance of starting under flooded or fouled conditions and less frequent tune-ups. Although gas mileage will be improved, the real value and cost justification is the reduction in maintenance costs. It is not unusual to go 50,000 miles between changes of plugs and breaker points. In most cases, that saves four tune-ups.

One other subtle advantage of a CD ignition system is that it draws around one ampere continuously rather than three to four amperes for conventional ignition systems, and five to seven amperes for transistor ignition systems. This savings in current could be important when driving in bad weather with most of the accessories operating. But more important, you have much more chance of starting with a weak battery. With most CD ignition systems, the spark output will be about the same from full battery voltage down to about 6 volts across a nominal 12-volt battery.

Because of the high output of the capacitive-discharge ignition system, the ignition cables, the rotor, and the distributor cap must be in good condition. A deteriorated ignition cable will not stand the hotter spark delivered by CD ignition. It will arc over and can cause misfiring. With an older car, the ideal solution is to install solid, metal, resistor-type ignition wire (called *magwire* in the trade). However, conventional TVRS resistor wiring in good condition is adequate.

Delta Mark Ten "B"

This inverter-type CD ignition system (Fig. 5-6) is marketed in kit and assembled form by Delta Products, Inc., by Radio Shack (under the *Archerkit* label), and by Heathkit. Components are installed on two printed-circuit cards fitted into an aluminum extrusion made weathertight by gasketed end plates as shown in Fig. 5-7.

While similar to the basic inverter-type CD ignition system described earlier, the Delta Mark Ten "B" includes several circuit features deserving comment. Referring to the schematic in Fig. 5-8, note that switching transistor Q3 is included in the gate-trigger circuitry of the SCR.

When the breaker points are closed, the junction of R5, R7, and R10 is pulled to ground. Transistor Q3 is not conducting and the

Fig. 5-6. Delta Mark Ten "B" capacitive-discharge ignition system.

SCR gate is effectively grounded through R10. During this period, capacitor C2 is charged to the level of the rectified output from the inverter. However, as the points open, a positive pulse suddenly appears at the junction of R10 and R11. This positive pulse, which is filtered by the low-pass network consisting of C6, R12, and C7, is coupled to the SCR gate, causing the SCR to conduct. Meanwhile,

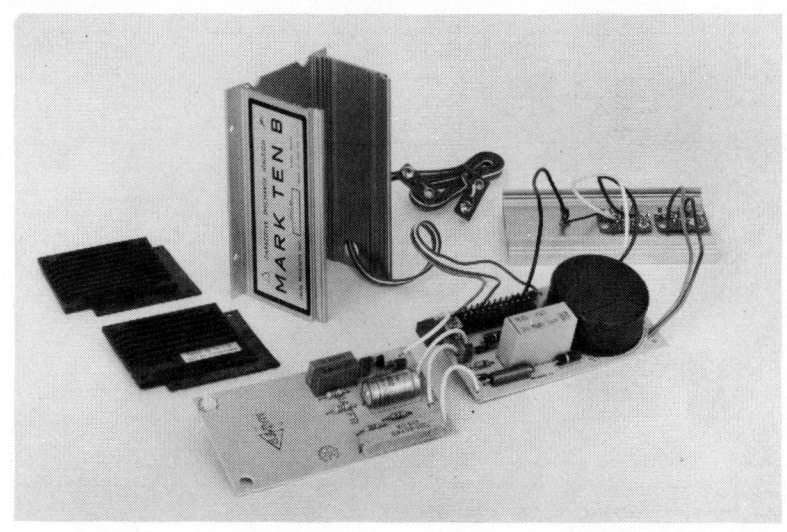

Fig. 5-7. Internal construction of Delta Mark Ten "B."

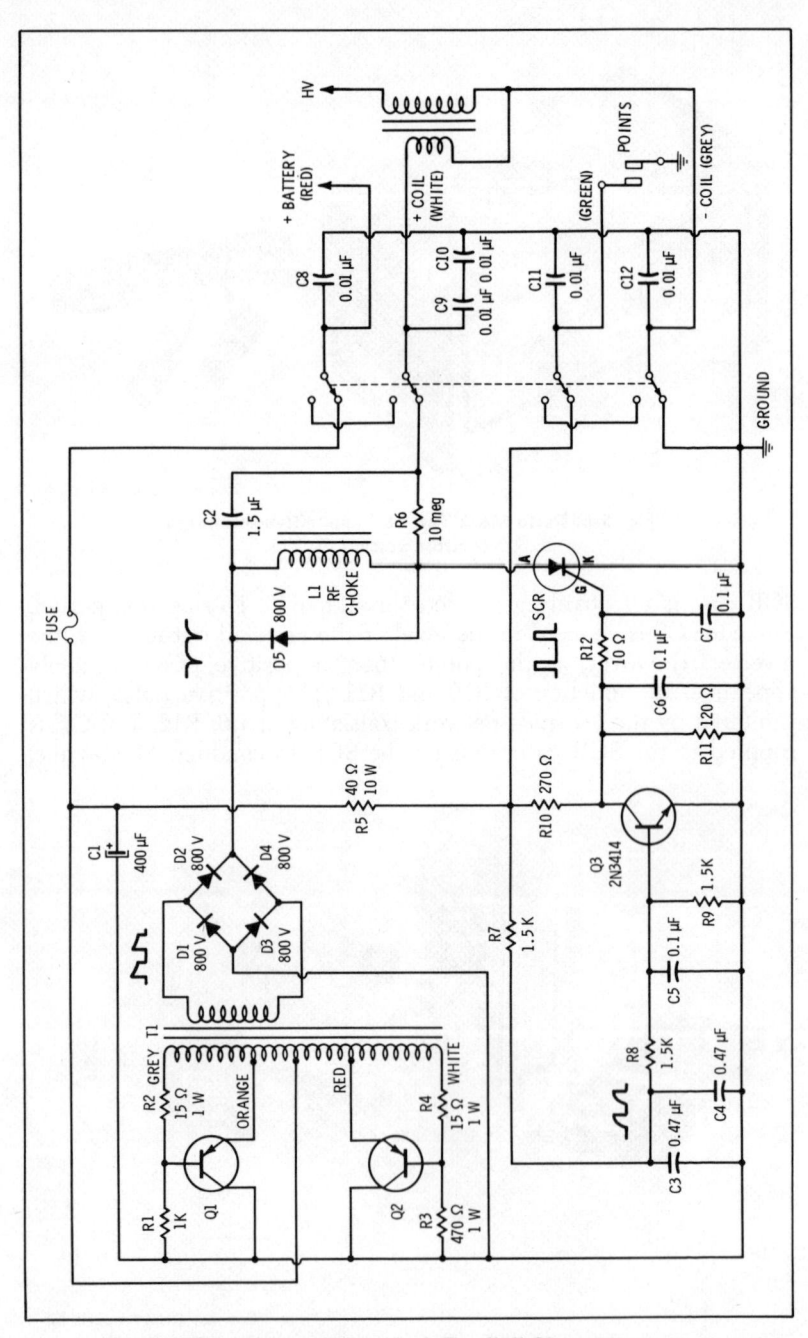

Fig. 5-8. Schematic of Delta Mark Ten "B" CD ignition system.

however, forward base bias is applied to Q3 through a delay network consisting of R7, C3 and C4. Thus, a brief time after the SCR conducts, Q3 saturates and pulls down the gate of the SCR. The cycle repeats as the points close.

The purpose of the switching transistor in the gate circuit is prevention of "false-triggering" and assurance that the SCR will not be turned on a second time after it has dumped a charge and turned off while the points remain open. This makes the circuit relatively immune to contact "bounce" and reduces the possibility of misfiring, which would adversely affect emissions rating of the engine. Energy delivery of the Mark Ten "B" is .086 joule with a spark duration of 250 microseconds.

Tri-Star "Tiger 500" CD Ignition System

Simplicity and ruggedness characterize the design of the "Tiger 500" capacitive-discharge ignition system shown in Fig. 5-9. The

Courtesy Tri-Star Corporation
Fig. 5-9. Tri-Star "Tiger 500" CD ignition system.

inverter employs a ferrite-core transformer and operates at a frequency of 8000 Hz, which is quite high in contrast with the more usual 500 to 1000 Hz of other units. The object of this design is to reduce the coil windings (and, hence, circuit losses) to a minimum and to prevent a phenomenon, common to early CD ignition designs, called "sync miss." In brief, "sync miss" results when an engine rate corresponds to the switching frequency (or its harmonics) of the inverter. The spark demand then occurs at a poor time in the charging curve of the storage capacitor, with the result that the engine misfires. The inverter frequency of the "Tiger 500" far exceeds even extreme engine speeds, eliminating this annoyance.

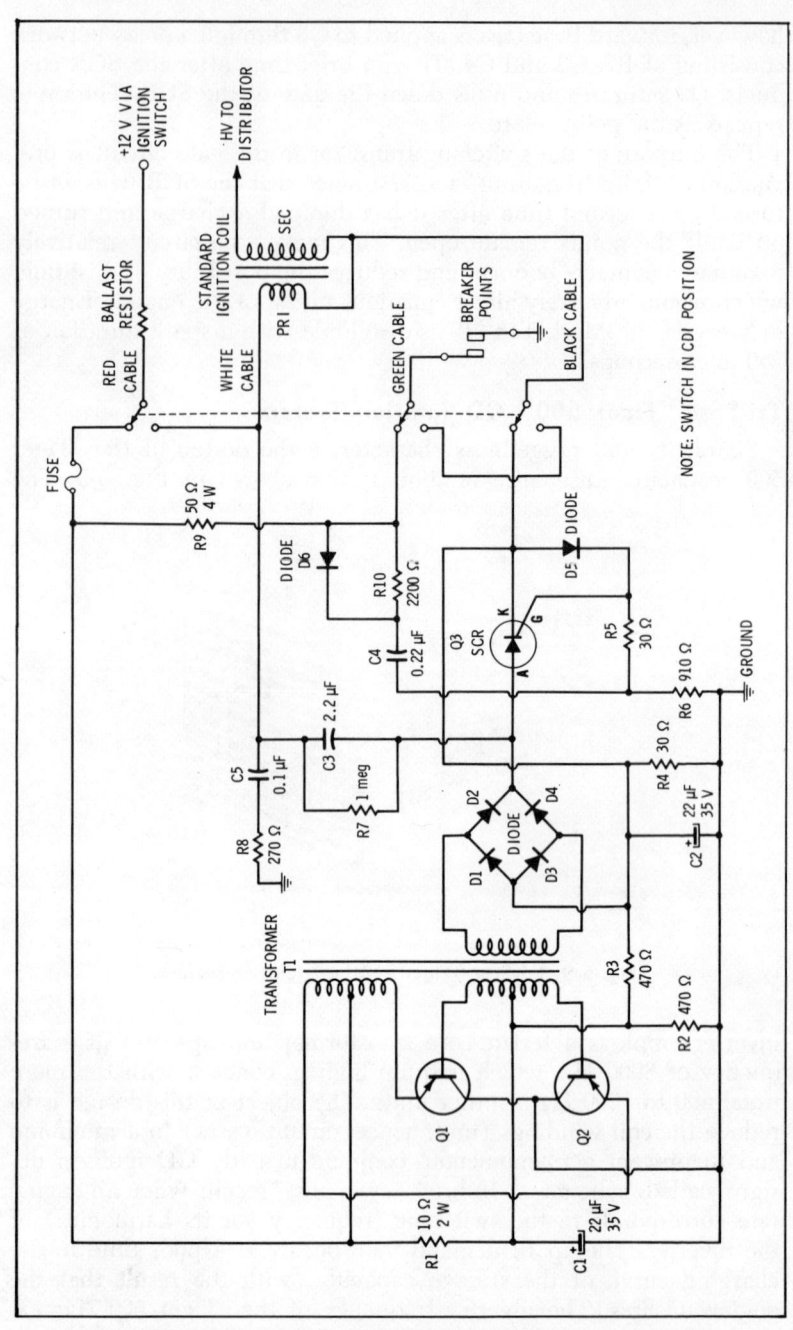

Fig. 5-10. Schematic of Tri-Star "Tiger 500" CD ignition system.

The schematic of the Tiger 500 is shown in Fig. 5-10. The inverter transistors, Q1 and Q2, are emitter coupled to a tapped winding on transformer T1, and positive feedback to sustain saturated switching is obtained from a second winding, appropriately phased. The collectors are grounded to a metal plate to effect heat sinking.

Assuming that the points are closed, the 8-kHz switching of the primary current produces a secondary voltage of 300 volts rms which is rectified by the diode bridge and applied to one plate of energy-storage capacitor C3. The opposite plate returns to ground through the ignition-coil primary, diode D5, and resistors R5 and R6 to complete the charging path. Meanwhile, current from the 12-volt supply flows through R9 and the breaker points to ground.

As the points open, a current path is created through R9 and diode D6, coupling a positive pulse through capacitor C4 and resistor R5 to the gate of SCR Q3, triggering it into the on state. Immediately, Q3 supplies a discharge path for capacitor C3 through the primary of the ignition coil. The large-scale discharge energy pulse is transformed by the conventional ignition coil to a peak spark voltage in the range of 40 to 45 kilovolts. Rise time of the discharge pulse is typically 2.5 microseconds, with approximately 150 millijoules of energy delivered to the primary. The period of the spark discharge is 350 microseconds. Commutation (turn-off) of the SCR is effected by collapse of the flux across the coil primary which generates a negative "undershoot" voltage that charges C3. When the negative voltage across C3 can no longer increase, the capacitor attempts to discharge in the reverse direction through the SCR. Instantly, the reverse-biased SCR assumes the blocking state and the inverter commences to recharge C3 during the recycle interval between spark demands. The unit is equipped with a switch to permit instant change from CD ignition to conventional ignition for tune-up or if a circuit failure should occur.

Cragar "Power Pack" CD Ignition System

The Cragar ignition system is shown in Fig. 5-11. This circuit (Fig. 5-12) utilizes a straightforward inverter circuit in which Q1 and Q2 switch power to the primary of T1. Unlike many other designs, however, the Cragar inverter incorporates zener diodes to protect the switching transistors from transient voltage spikes. This eliminates a potential failure source in CD ignition system operation.

The trigger circuitry of the Cragar ignition system is also unusual in that the trigger pulse to the SCR is negative and, therefore, is applied to the cathode rather than the gate. In operation, closure of the breaker points grounds the junction of R8 and R9, reverse biasing transistor Q3 and holding it off. A positive potential is applied to the cathode of the SCR, causing it to remain off. During this period,

Fig. 5-11. Cragar "Power-Pack" CD ignition system.

the rectified output from the inverter charges C1 through the primary of the ignition coil. As the breaker points open, Q3 is forward biased and provides a path for charging current to flow into C3. Instantly, the cathode of the SCR is pulled negative, effectively pulsing the grounded gate positive. The SCR conducts, simultaneously stopping the inverter and completing the discharge path for capaci-

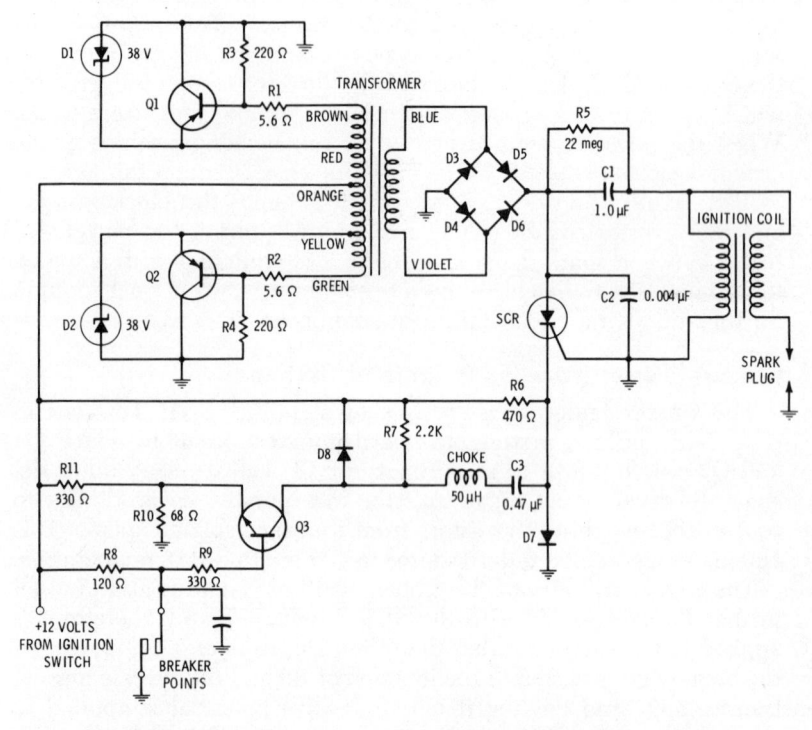

Fig. 5-12. Schematic of Cragar "Power-Pack" CD ignition system.

82

tor C1 through the coil primary. The cycle repeats as the SCR commutates.

With an energy output of 0.12 joule and a spark duration of 260 microseconds, the Cragar design capably fires even badly fouled and worn plugs. It has a minor disadvantage in that a switch is not provided for changing from CD to conventional ignition. This means wire-swapping in the event of failure, or when tuning up with a standard tach/dwell meter.

Delco-Remy Capacitive-Discharge System

The Delco-Remy capacitor-discharge ignition system features a control unit and a special ignition coil that are used with the conventional distributor. Fig. 5-13 is a wiring diagram showing the basic electrical connections for this system. This system is designed to be used only on vehicles employing a 12-volt battery with negative ground.

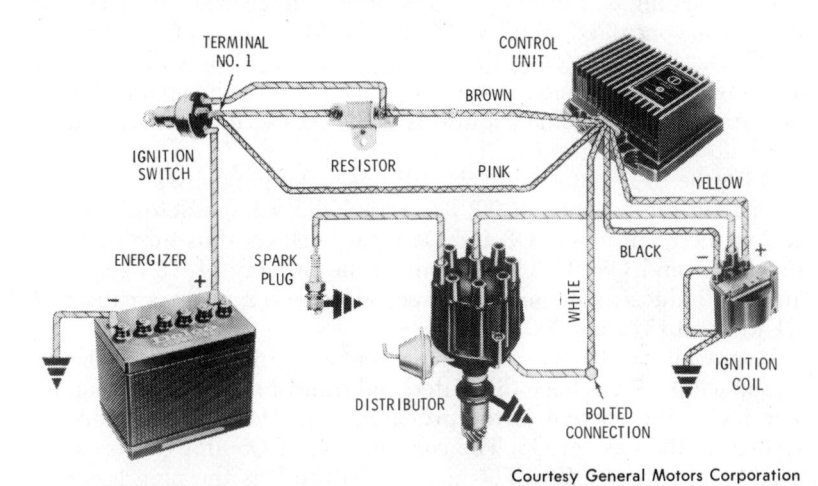

Courtesy General Motors Corporation

Fig. 5-13. Electrical connections for Delco-Remy CD ignition system.

The special ignition coil and control unit are designed to operate together, and are claimed to provide superior ignition performance under all operating conditions. Four important advantages attributed to this capacitor-discharge system are lower contact-point current with extended contact life, superior starting ability particularly with wet plugs, a higher secondary voltage throughout the engine-speed range, and extended spark plug life due to improved ability to fire wet or fouled plugs. The only alleged maintenance requirements are those procedures applicable to the conventional distributor and spark plugs.

The schematic for the Delco-Remy CD ignition system is shown in Fig. 5-14. With the distributor contacts closed, capacitor C7 and capacitor C1 remain charged from a previous cycle. The voltage potential across resistor R17 is approximately equal to the system voltage, since the primary resistance of T2 is very small. It is the voltage that causes transistors Q3 and Q4 to be biased on.

As the distributor contacts open, transistor Q1 turns on and capacitor C7 discharges through Q1, the energizer, and R3. This causes transistor Q2 to turn off which decreases the current in the primary of T1 and imposes a positive voltage at the base of Q5, initiating the turn on of transistors Q5 and Q6. Forward bias for Q5 is sustained by current flow through R5, R14, and diode D4. When Q5 and Q6 initially turn on, the potential across resistor R17 drops almost to zero, causing transistors Q3 and Q4 to turn off—a condition necessary for transistors Q5 and Q6 to be on.

The increasing current in the primary transformer T2 causes the trigger winding of T2 to impose a positive voltage at the gate (G) of the silicon controlled rectifier (SCR). The SCR turns on, and capacitor C1 discharges through the primary of the ignition coil, inducing a voltage in the secondary winding which causes the spark plug to fire. Actually, the voltage pulse is induced twice, as described in the following paragraphs.

With the distributor contacts still open, the previously increasing current in the primary of T2 has raised the voltage across resistor R17 to the point where Q3 and Q4 turns back on, thus turning transistors Q5 and Q6 off. The current in the primary of T2 decreases, and the induced voltage in the secondary of T2 charges capacitor C1 to approximately 300 volts.

Meanwhile, with the distributor contacts still open, the charge on capacitor C7 has been dissipated and transistor Q2 turns on again. The increasing current in the primary of transformer T1 induces a voltage at the base of Q5. The combination of Q5 and Q6 turns on, and the plug fires again. This unique feature fires the plug twice in rapid succession to ignite the air/fuel mixture. After the second firing, the system reverts to initial status, during which time capacitor C1 is charged again and is ready for the next spark plug firing. The distributor contacts then close, and the previous cycle just described then repeats.

The other control-unit components not yet mentioned refine operation of the system. The SCR is protected from high reverse voltages and transient voltages by C2 and R22 when reverse current flows in the ignition-coil primary with a secondary lead, such as a spark plug lead, open. Resistors R19, R21, and R25 have very high resistance values and allow the charge on C1 to bleed off when the system is not in operation. Zener diode D5 prevents excessive volt-

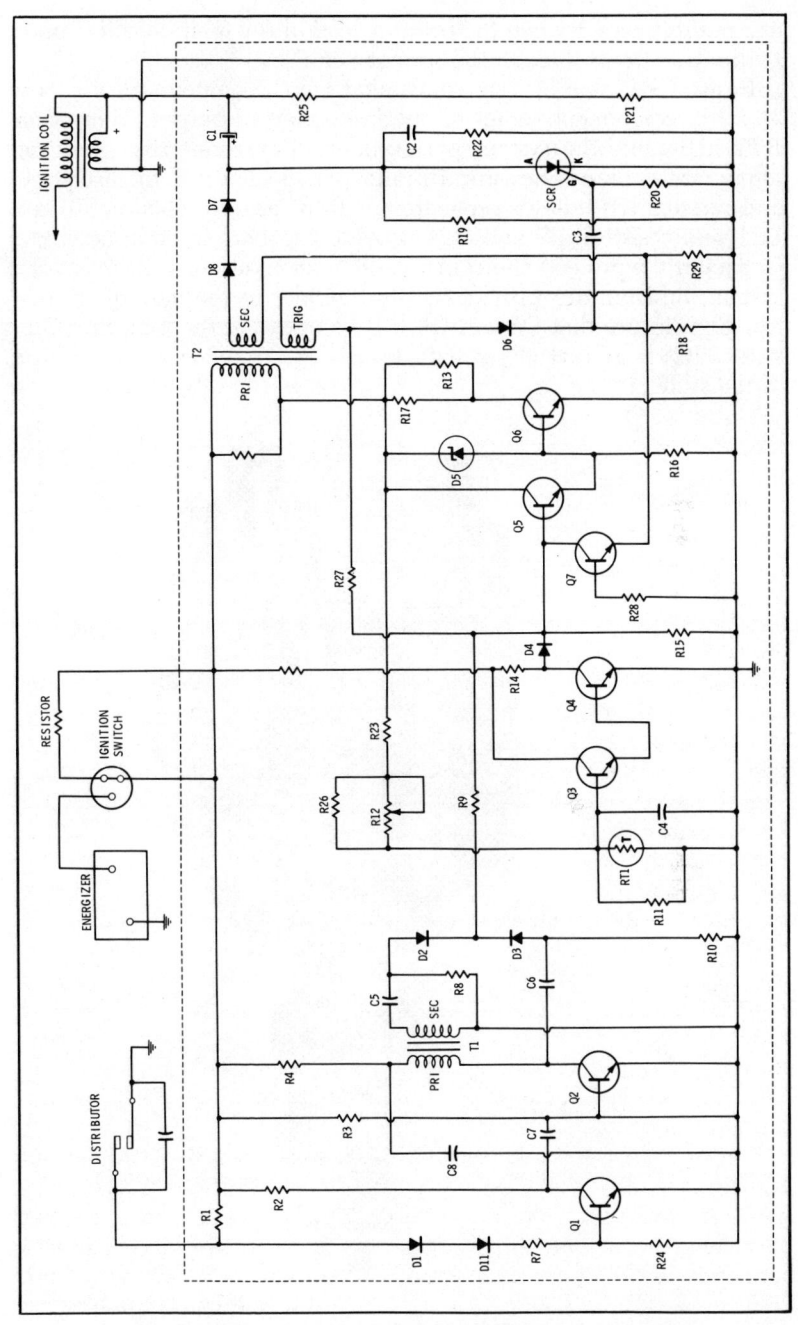

Fig. 5-14. Schematic for Delco-Remy CD ignition system.

age buildup on C1 when the primary lead of the ignition coil is open by limiting the voltage in the primary of T2.

Resistor R12 is a factory adjustment which compensates for variations in component values to obtain a specified output. Thermistor RT1 maintains the system performance throughout the operating temperature range. Resistor R8 allows capacitor C5 to discharge, and resistor R10 allows capacitor C6 to discharge. Diodes D1 and D11 and resistors R7 and R24 provide the bias for transistor Q1. Capacitor C8 protects the control unit from transient voltages in the system. Resistor R27 provides a positive bias at the base of transistor Q5 to insure that Q5 and Q6 will be on when the plug fires. Resistors R28 and R29 along with transistor Q7 refine the operation of the system.

Courtesy General Motors Corporation

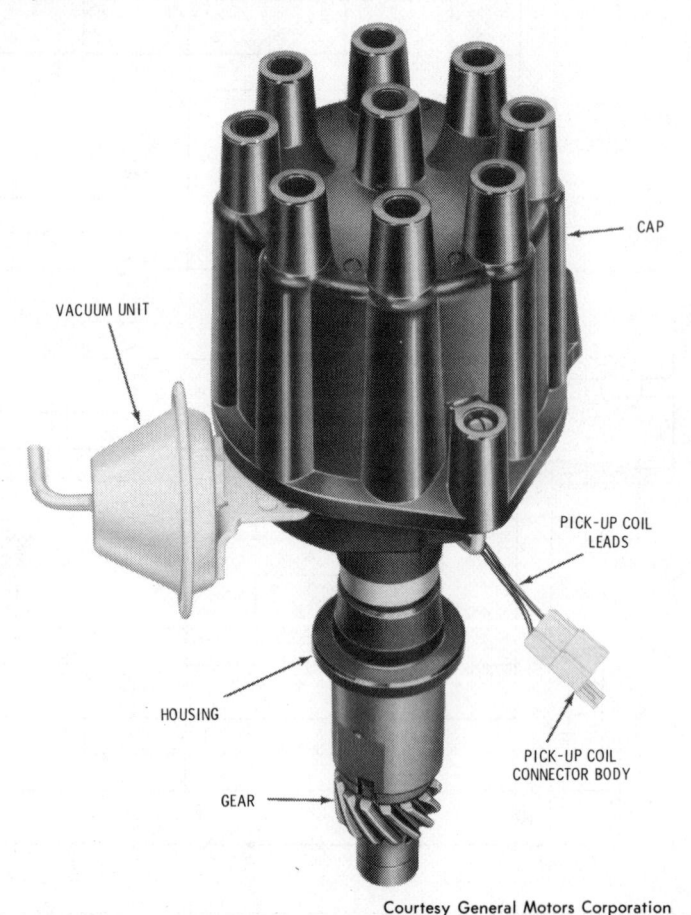

Fig. 5-15. Typical magnetic pulse distributor.

Delco-Remy Magnetically Triggered CD Ignition System

This magnetically triggered capacitive-discharge ignition system features a specially designed pulse distributor, an ignition pulse amplifier, and a special ignition coil. The system operates to charge a capacitor to a high voltage between spark plug firings. On a signal from the distributor, the capacitor is then discharged through the primary of the ignition coil to fire the spark plug.

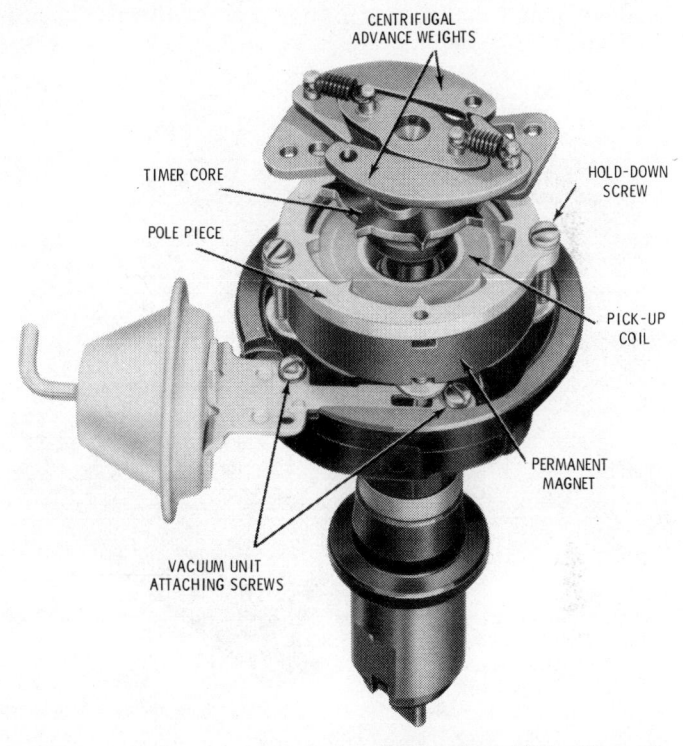

Courtesy General Motors Corporation

Fig. 5-16. Partially exploded view of magnetic pulse distributor with cap removed.

A typical magnetically triggered distributor is shown in Fig. 5-15. Although the external appearance of the unit resembles a standard distributor, the internal construction is quite different. As shown in the partially exploded view of Fig. 5-16, an iron timer core replaces the conventional breaker cam. The timer core has the same number of equally spaced projections, or vanes, as the engine has cylinders. The timer core rotates inside a magnetic pickup assembly which replaces the conventional breaker plate, contact point set, and condenser assembly.

The magnetic pickup assembly consists of a ceramic permanent magnet, a pole piece, and a pickup coil. The pole piece is a metal plate having a number of equally spaced internal teeth; that is, one tooth for each cylinder of the engine. The magnetic pickup assembly is mounted over the main bearing of the distributor housing and is made to rotate by the vacuum control unit, thus providing vacuum advance. The timer core is rotated in reference to the shaft by conventional advance weights, thus providing centrifugal advance.

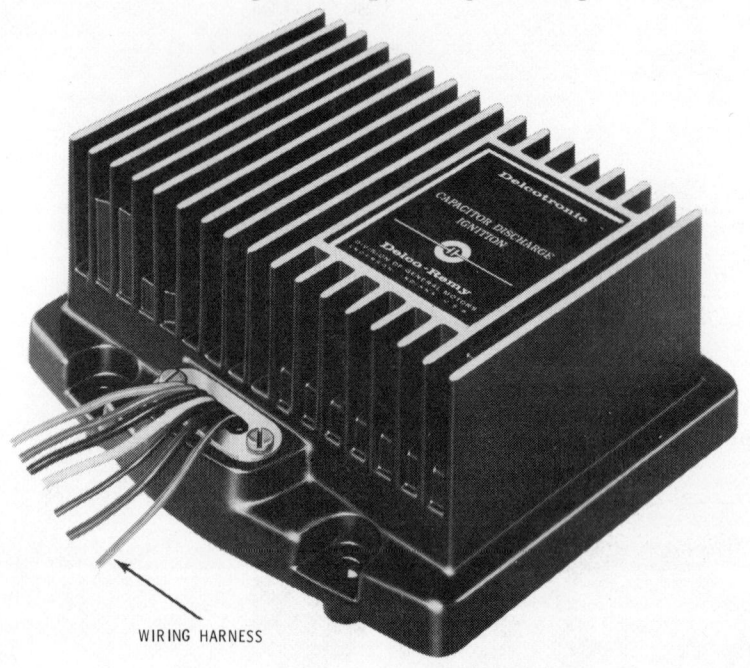

WIRING HARNESS

Courtesy General Motors Corporation

Fig. 5-17. Typical ignition-pulse amplifier.

The ignition-pulse amplifier is shown in Fig. 5-17. This unit consists primarily of transistors, diodes, resistors, capacitors, a thyristor, and a transformer mounted on a printed-circuit board. The schematic for the pulse amplifier is shown in Fig. 5-18.

Resistor R4, transistor Q2, transformer T1, and the four-diode bridge act together as an inverter/rectifier to charge capacitor C1 to approximately 300 volts dc. The operation of the inverter circuit is a departure from most designs in present use. A forward-bias voltage is applied to the base-emitter junction of transistor Q2 by the battery through resistor R4, turning Q2 partially on. An increasing

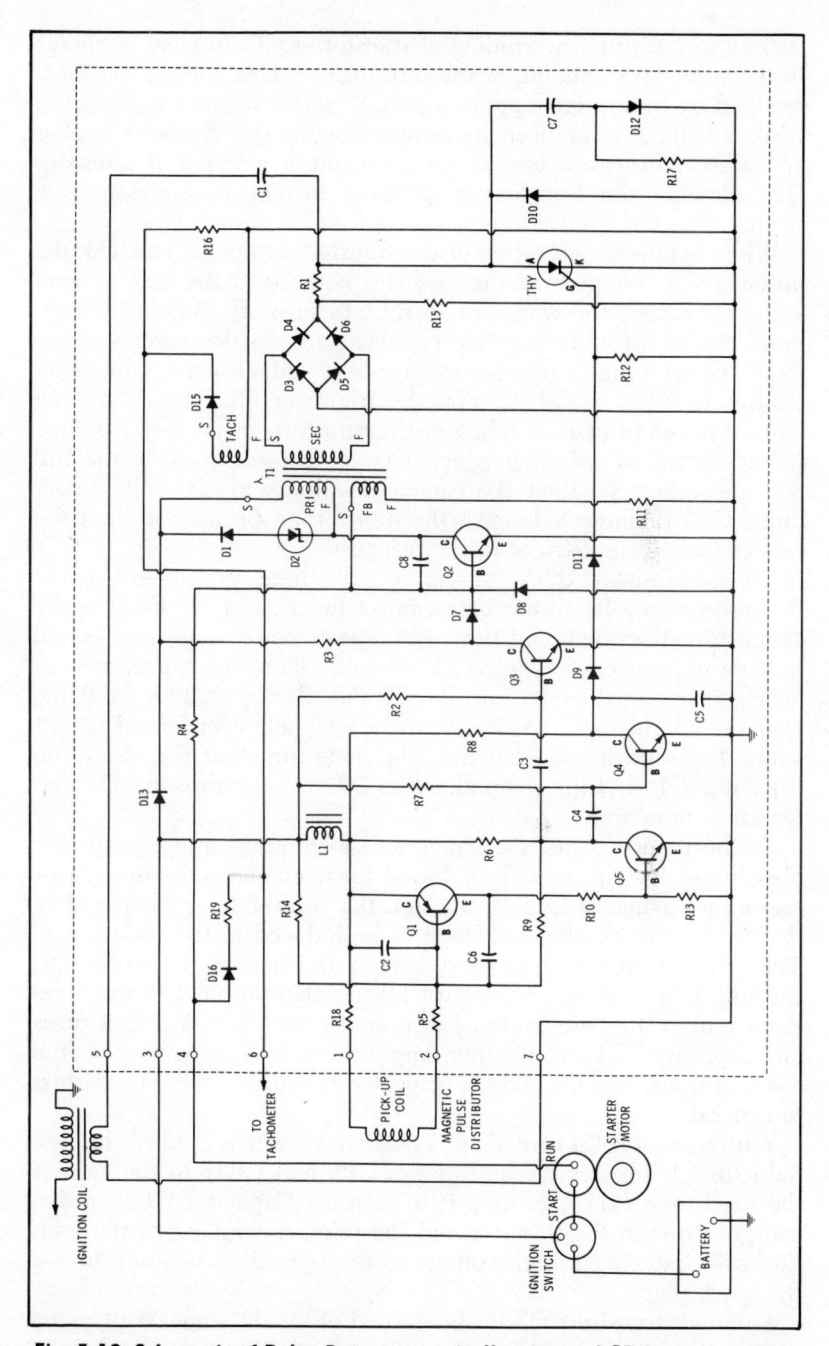

Fig. 5-18. Schematic of Delco-Remy magnetically triggered CD ignition system.

current in the primary winding of transformer T1 induces a voltage in the secondary winding of the transformer. This voltage, which is rectified by the bridge, applies a partial charge to the capacitor C1. Also, a voltage is induced in feedback winding FB which applies an additional forward bias to the base-emitter junction of transistor Q2, allowing the transformer primary current to increase even further.

When magnetic saturation of the transformer core is reached, the absence of a voltage from the feedback winding at the base of transistor Q2 causes the transistor to start to turn off. When this happens, the transformer primary current starts to decrease, and the feedback winding imposes a reverse-bias voltage across the base-emitter junction of Q2, turning the transistor off. The decreasing current in the primary winding of the transformer induces a voltage in the secondary winding which charges capacitor C1 to the full 300-volt value. To limit the capacitor voltage to 300 volts, zener diode CR2 is connected across the transformer primary to limit the rate of decreasing current in the primary.

When transistor Q2 is turned off, the battery voltage applied through resistor R4 to the base-emitter junction of Q2 initiates the action just described, and the cycle repeats many times per second to keep the capacitor charged to 300 volts. Thus, the capacitor voltage is maintained at its maximum value during engine cranking, even though the battery voltage may be well below its normal 12-volt value. It should be noted that at the same time that the charge on capacitor C1 is being maintained at 300 volts, transistors Q3 and Q4 are conducting.

As the engine turns, the vanes on the rotating iron core in the distributor line up with the internal teeth on the pole piece. This establishes a magnetic path through the core of the pickup coil in the distributor, causing a voltage to be induced in the pickup coil. This voltage applies a forward bias to the base of transistor Q1, causing it to turn on. A forward bias is then applied through resistor R10 to the base-emitter junction of transistor Q5, and it turns on. Capacitors C3 and C4 then apply a reverse-bias voltage to the bases of transistors Q3 and Q4 respectively, causing these transistors to turn off.

With transistor Q4 turned off, a positive voltage is applied through inductor L1, resistor R8, and diodes CR9 and CR11 to the gate of the thyristor (THY), causing it to turn on. Capacitor C1 then discharges through the thyristor and the primary winding of the ignition coil, inducing a high voltage in the secondary winding to fire the spark plug.

Although transistor Q3 is also turned off at this time, it plays no part in the circuit operation for as long as the ignition switch is in the crank position. When the engine is being cranked, the action of transistor Q2 is initiated by battery voltage applied through resistor R4, and the oscillator operation continues to keep capacitor C1 charged as previously described. Transistor Q3 does not become effective until the engine is running.

When a spark plug has fired and the engine has started, the ignition switch is returned to the run position and the No. 4 terminal on the pulse amplifier is no longer energized. Although the system no longer operates as a free-running oscillator with the No. 4 terminal de-energized, the oscillator action continues long enough after the plug fires to charge capacitor C1 to the 300-volt value. The capacitor therefore is charged and ready to fire the next spark plug.

With the engine running and no pulse from the distributor, the charge on capacitors C3 and C4 has been dissipated and transistors Q3 and Q4 have turned back on. At the same time, transistor Q2 and transformer T1 are inoperative. Transistor Q2 is turned off because transistor Q3 in the on condition allows battery current through resistor R3 to pass directly to ground, bypassing diode CR7 and the base-emitter junction of Q2. It can now be seen that with the engine running and the ignition switch in the run position, transistor Q3 keys the action of transistor Q2 and the system will not operate as a free-running oscillator.

With the engine running and a voltage induced in the distributor pickup coil, transistors Q1 and Q5 turn on in the same manner as during cranking. Transistor Q4 is turned off by a pulse through capacitor C4 and a positive voltage is applied to the gate of the thyristor (THY). The thyristor turns on, and capacitor C1 discharges through it and the primary winding of the ignition coil to fire the spark plug in the same manner as during cranking.

Due to the inductive-capacitive effect of the ignition-coil primary and capacitor C1, the direction of current flow through the coil will reverse, and will enable diode CR10 to assist the transformer in charging capacitor C1 for the next plug firing. Also, the current through diode CR10 imposes a reverse-bias voltage across the thyristor (THY), causing it to turn off. When the current flow reverses again, a very high value of forward-bias voltage due to the recovery effect of diode CR10 may be impressed across the thyristor. This high forward-bias voltage would cause the thyristor to turn on without a signal at the gate. To prevent this unwanted occurrence, capacitor C7 and diode CR12 reduce the rate of voltage increase across the thyristor. Resistor R17 reduces the rate of voltage change imposed on the thyristor when capacitor C7 discharges.

Since transistor Q3 has been turned off by a pulse through capacitor C3, a forward-bias voltage through resistor R3 and diode CR7 initiates the turn on of transistor Q2. An additional forward-bias voltage is applied to the base-emitter junction of transistor Q2 by the feedback winding FB, and the transformer primary current increases to its maximum value to partially charge capacitor C1. When the maximum primary current is reached, the loss of voltage from the feedback winding causes transistor Q2 to turn off. The transformer primary current decreases, and the transformer secondary charges capacitor C1 to the full 300-volt value. The capacitor is now ready to fire the next plug. Inductor L1 acts to prevent induced voltages in the primary of transformer T1, as well as transient voltages from the cranking and accessory circuits, from being impressed into the triggering circuit consisting of transistors Q1, Q3, Q4, and Q5 and their associated components.

Thermo King "Thermotronic" Magnetically Triggered CD Ignition System

The engines in fleets of taxis, trucks, police cars, and industrial plant vehicles (as well as stationary engines that power pumps and generators) present special maintenance problems to their owners. While an average driver rolls up 10,000 engine miles per year, these hard-working power sources work day and night, often running 100,000 miles each year. The Thermo King Corporation contends that hard use like this erases many of the benefits of CD ignition if the system is triggered by wearout-prone conventional breaker points. Therefore, the Thermotronic CD ignition system eliminates the points and substitutes a magnetic triggering unit.

Components of the Thermotronic system are shown in Fig. 5-19. The trigger unit mounts in the distributor on a base plate, shaped like that of a point set. The wiring harness connects it to the converter unit which is mounted on the firewall or fender well of the vehicle. Fig. 5-20 illustrates the hookup of the system in a vehicle.

The magnetic trigger unit is positioned so as to sense the rotation of the conventional distributor cam. The magnetic flux through the

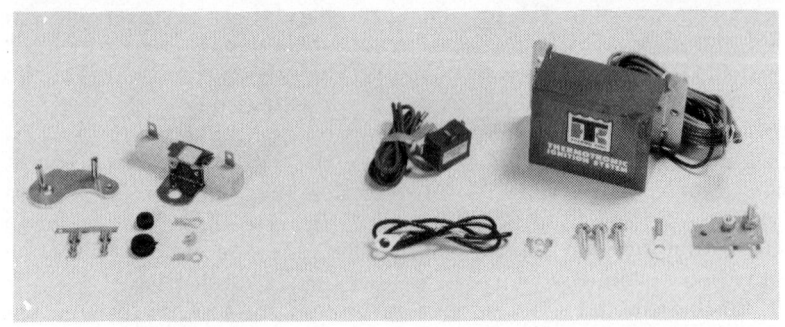

Courtesy Thermo King Corporation

Fig. 5-19. Components of the Thermo King CD ignition system.

coil in the trigger unit increases as a cam high point passes the unit pole piece and decreases as a flat lies opposite the pole piece. This rise and fall in flux density results in generation of a rising and falling potential in the coil. Amplified and shaped by the integral electronics of the trigger unit, this potential becomes a switching signal that keys the converter unit.

Courtesy Thermo King Corporation

Fig. 5-20. Thermo King CD ignition system installed in vehicle.

Fig. 5-21 is a schematic of the system electronics. T1 is the transducer coil, and its output is applied to operational amplifier A1. Feedback for stability is provided by capacitor C1, so that higher-frequency components in the amplified coil signal are suppressed. The amplified output of A1 drives cascaded switching amplifiers Q1 and Q2. Q1 is normally off and Q2 is normally on when the output from A1 is low (pole piece facing a distributor cam flat). However, as the output from A1 rises, Q1 switches on, reverse biasing Q2. (Q2 receives its collector current from the converter unit.) As Q2 switches off, its collector rises immediately to +12 volts, providing a step trigger signal to the converter unit. This step lasts until the output from A1 has fallen off and transistors Q1 and Q2 reverse their state.

In the converter unit, high-frequency oscillator transistor Q5 switches power to the primary of T1, which is stepped up to approximately 250 volts and rectified by diode D3. The half-wave rec-

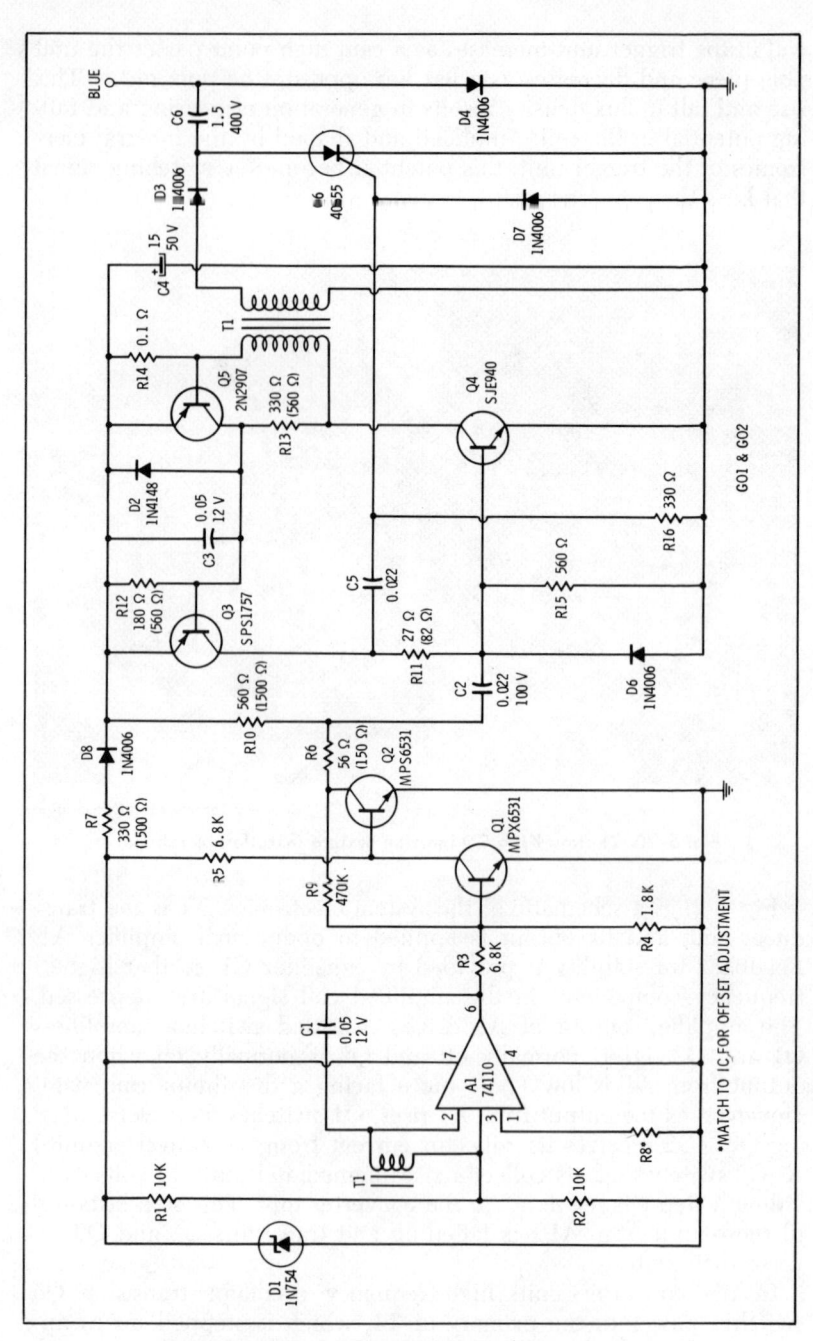

Fig. 5-21. Schematic of Thermo King ignition system.

tified pulses charge energy-storage capacitor C6 through a return path provided by the ignition coil primary during the interval.

Assume now that a positive trigger pulse is applied from the trigger unit through C2 to the base of transistor Q4. Q4 immediately switches on, momentarily removing collector feedback energy from Q5. Instantly, oscillation ceases and transistor Q3 is simultaneously forward biased, supplying a large positive current pulse through C5 to the gate of SCR Q6. This pulse triggers and latches SCR Q6 into conduction, discharging storage capacitor C6 through the primary of the ignition coil to provide spark potential. The energy collapse in the ignition coil reverses the polarity applied to the SCR, causing it to commutate. Meanwhile, the cam high point has passed the pole piece of the trigger unit and no trigger input signal is applied to the converter unit. Transistors Q3 and Q4 are reverse biased and transistor Q5 commences oscillation, beginning another cycle.

Providing an output of approximately 68 millijoules over an operating temperature range of −40° to +230°F, the Thermotronic system reliably fires plugs having gaps of up to .050 inch in a 135-psi environment. Installation kits fitting any standard distributor are available from the manufacturer.

Jermyn CD Ignition System Mk. 2

Readers in the United Kingdom may find the Jermyn Mk. 2 CD system an economical answer to converting over to capacitive-discharge ignition. Based on construction details originally published in *Practical Wireless*, the system is available for both positive-ground and negative-ground electrical systems.

Fig. 5-22 illustrates the two versions of the Jermyn CD ignition system. In both versions, the inverter and bridge rectifier circuit is conventional. The output voltage is higher than normal (450 volts), but storage capacitor C1 is somewhat smaller than usual (0.47 μF). Thus, the resulting output of approximately 90 millijoules is in the average range for most CD ignition systems. Triggering of the SCR and commutation after discharge are conventional. Additional details regarding the kit for either version may be obtained from Jermyn Industries. Vestry Estates, Sevenoaks, Kent, England.

Sydmur CD Ignition Systems

Perhaps the most striking feature of the Sydmur CD ignition system shown in Fig. 5-23 is the small number of components. As shown in the schematic (Fig. 5-24), the Sydmur *Flyaway* unit employs a "brute-force" inverter, a conventional bridge rectifier, and a conventional SCR discharge circuit. Marketed by Sydmur Electronic Specialities of Brooklyn, New York, the system includes none of the

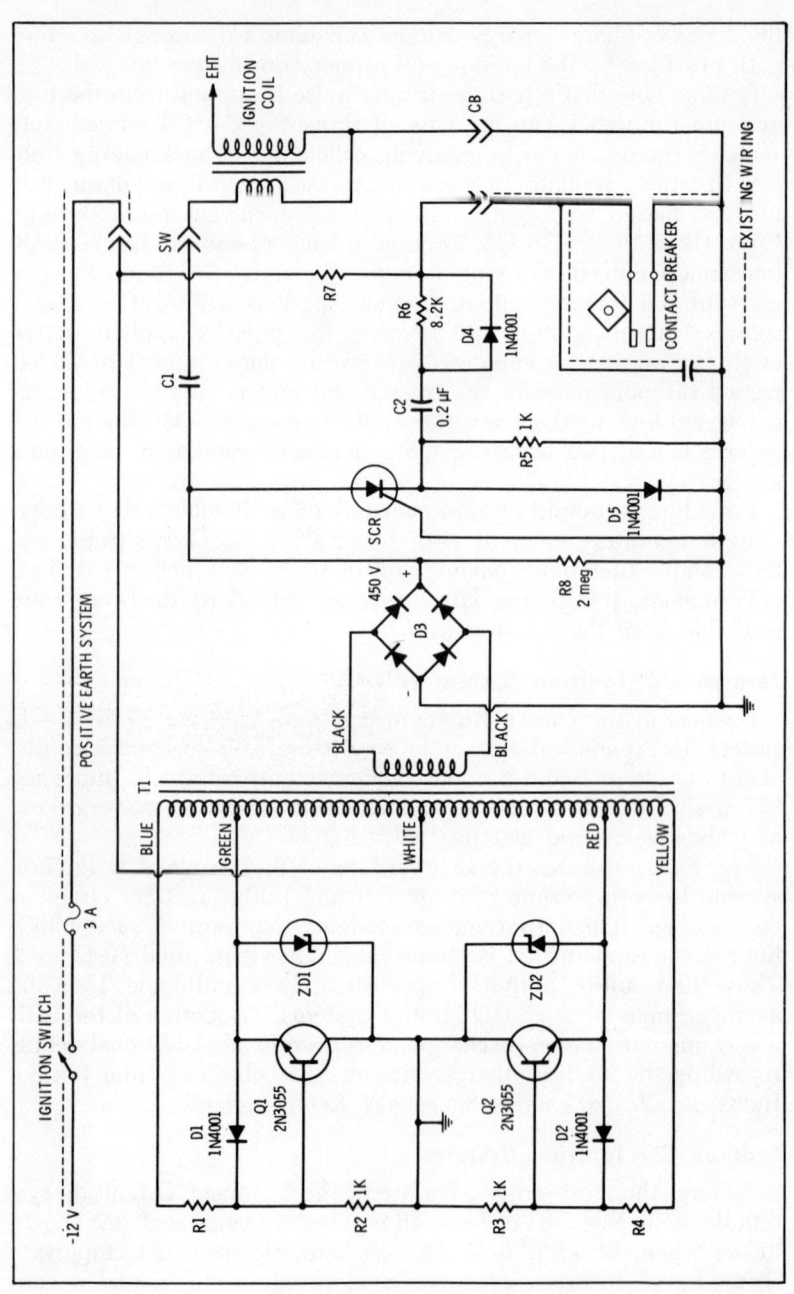

(A) Positive-ground system.

Fig. 5-22. Schematic of the Jermyn

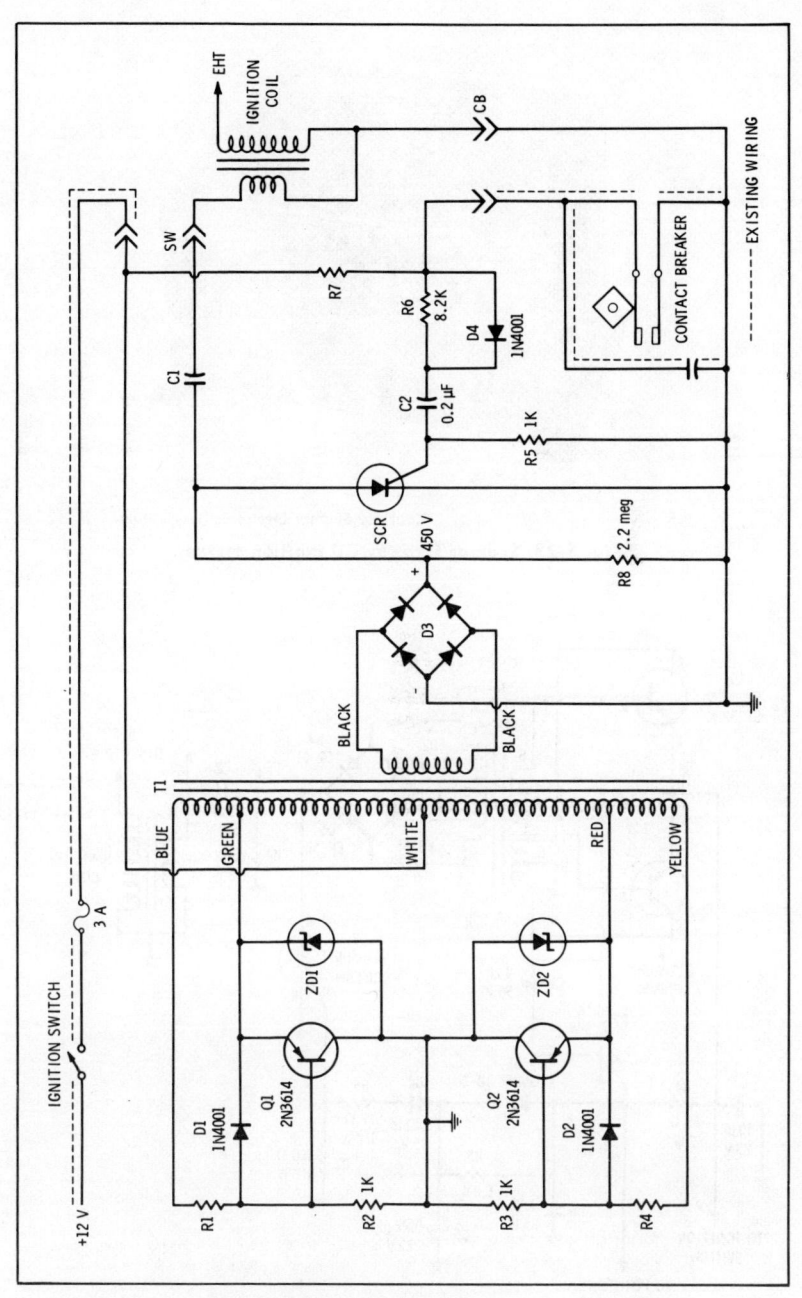

(B) *Negative-ground system.*

Mk. 2 CD ignition system.

Courtesy Sydmur Electronic Specialities

Fig. 5-23. Sydmur *Flyaway* CD ignition system.

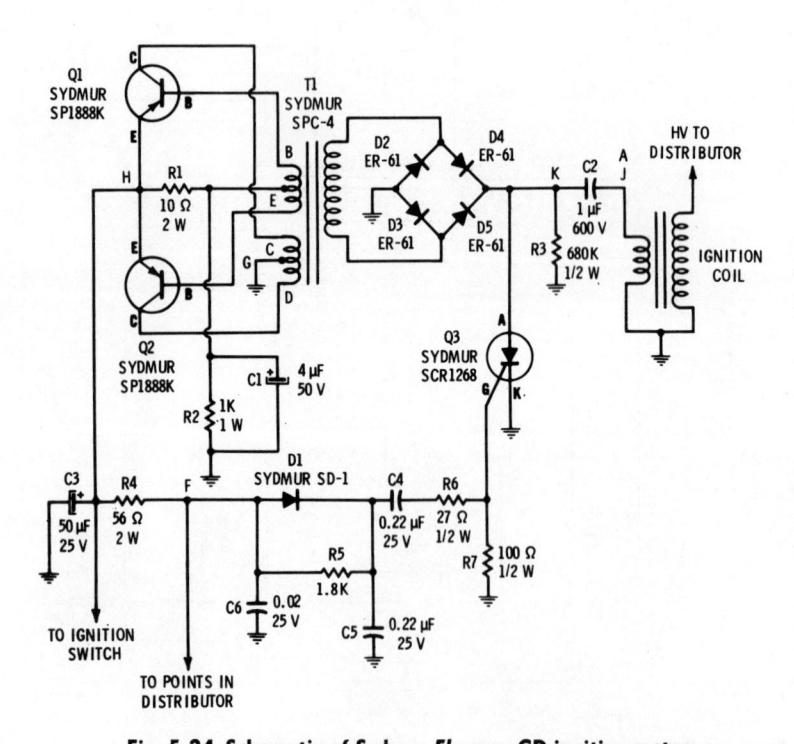

Fig. 5-24. Schematic of Sydmur *Flyaway* CD ignition system.

98

elaborate "single-trigger" and anti-sync-miss circuits employed in some competing units. The Sydmur system traces its origin to a construction article appearing in the June 1965 issue of *Popular Electronics* magazine.

The Sydmur *Compac* CD ignition system shown schematically in Fig. 5-25 employs a bare minimum of circuitry. The inverter circuit, comprised of a single transistor (Q1) operates as a blocking oscillator. Bias for Q1 is provided by divider network R1 and R2. Feedback to sustain oscillation is made positive by the phasing of the transformer windings connected to the emitter and base of Q1.

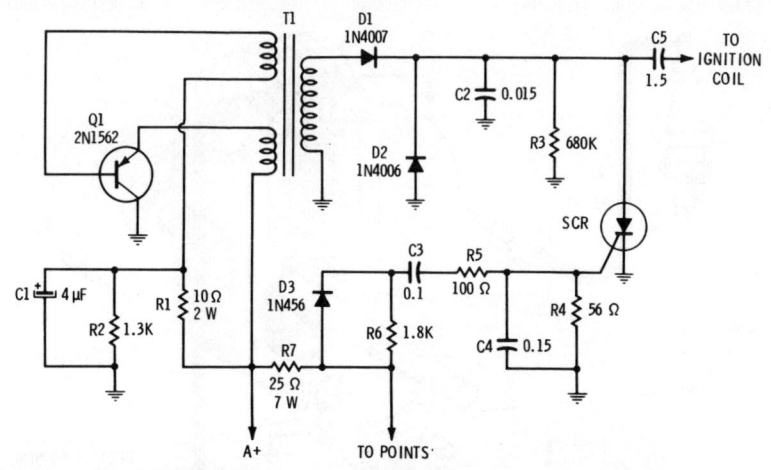

Fig. 5-25. Schematic of Sydmur *Compac* CD ignition system.

In operation, the flow of base current results in an increase in collector current, building a magnetic field of increasing strength about the windings of T1. Each flux increase is accompanied by an increase in base current, until Q1 reaches saturation and the collector current ceases. The collapsing field about T1 induces a potential across the secondary which is rectified and stored in C5. This occurs many hundreds of times between each firing of the SCR. Apart from the inverter, the operation of this CD ignition system is conventional.

MAGNETO-TYPE CD IGNITION SYSTEMS

Power mowers, chain saws, snow throwers, and snowmobiles rely upon effective ignition for their operation just as heavily as do automobiles. In fact, since most small-displacement engines used in these devices are of the one- or two-cylinder variety, operating with a two- or four-stroke cycle, there is little tolerance for misfiring. For this reason, there is an unprecedented rush in the power equipment and recreational vehicle fields to equip small engines with reliable *magneto-type CD ignition.*

Conventional Magneto-Ignition Operation

As shown in Fig. 5-26, the conventional magneto system employs a rotating magnet coupled to a pair of coil windings by a closed magnetic circuit formed by silicon-steel laminations. As the magnet rotates, its flux cuts both the primary and secondary windings, induc-

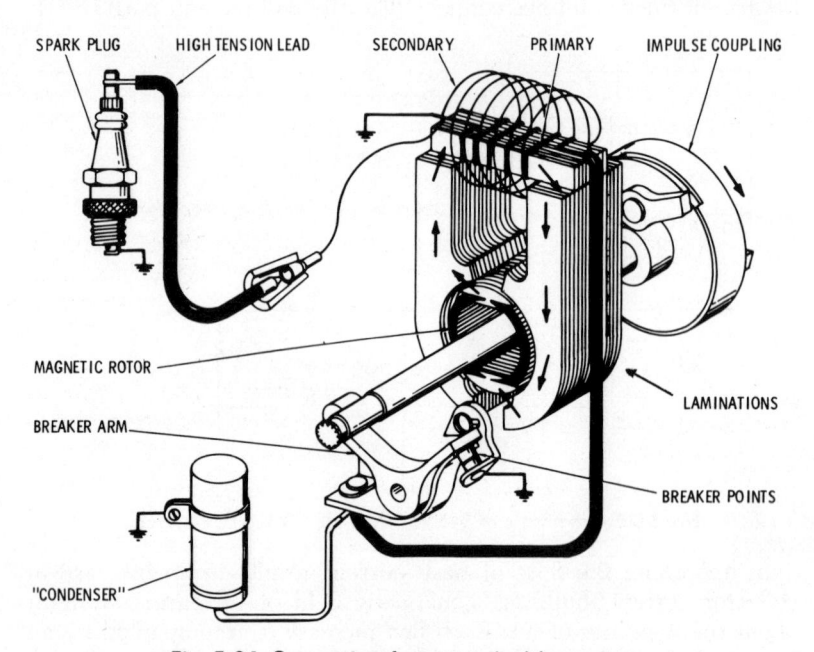

Fig. 5-26. Conventional magneto ignition system.

ing a current in each. As peak current occurs in the primary, a set of breaker points actuated by the magnet shaft opens, interrupting the primary current and causing the flux to collapse through the secondary. The collapse of flux induces a high voltage in the secondary winding which is impressed across the plug gap, causing the spark that ignites the fuel/air mix. The condenser performs the same function as its counterpart in the conventional Kettering ignition system. It produces rapid collapse of primary-circuit flux and reduces breaker-point arcing.

The failure-prone mechanism of the magneto ignition system is the same as that of the conventional Kettering ignition system—the breaker points. Inevitably, the contact resistance increases with age

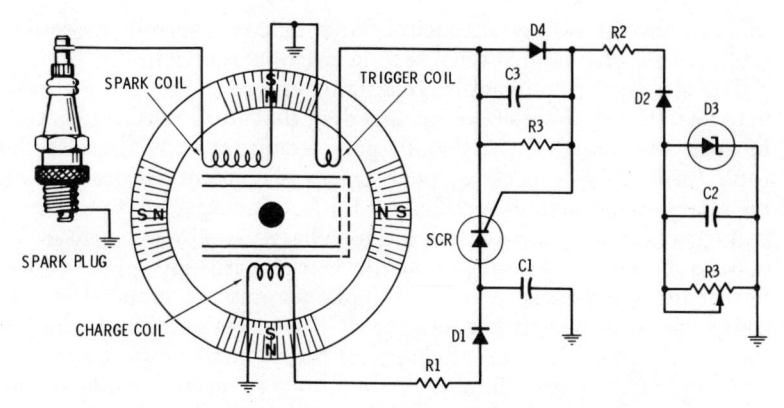

Fig. 5-27. Schematic diagram of magneto-type CD ignition system for small engines.

and arcing, and rubbing-block wear changes the critical relationship between the time that peak current flows in the primary winding relative to the time that points actually open. The result is occasional misfiring, then serious power loss, and finally a "no-start."

Magneto-Type CD Ignition Operation

To overcome the shortcomings of conventional magneto ignition, designers discarded the breaker points and added a third coil wind-

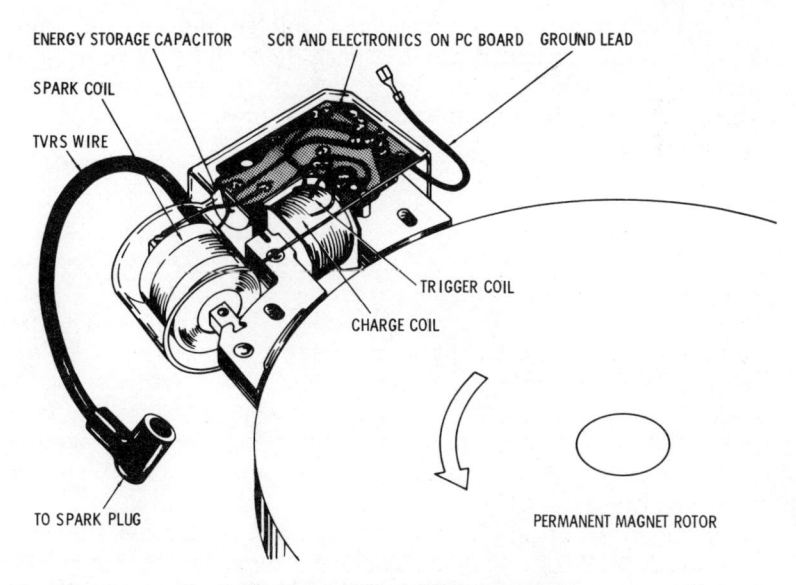

Fig. 5-28. Magneto-type CD ignition system.

ing, an energy-storage capacitor, and silicon controlled rectifier (SCR). The schematic of the resulting circuit is shown in Fig. 5-27.

The magnetic rotor was also redesigned to include four salient pole magnets. Three of these are arranged so that their north poles face inward. The fourth has its south pole facing inward. Thus, as the rotor turns, three successive peak pulses of current are induced in the charge winding and rectified by D1, producing a relatively high-voltage charge on capacitor C1. This charge is stored because the nonconducting SCR provides no discharge path. However, as the fourth magnet passes, a negative voltage pulse is induced in the trigger winding which triggers the SCR into conduction. Instantly, the SCR "dumps" the stored charge of the capacitor into the trigger coil, inducing a considerably higher voltage in the magnetically coupled spark coil, which creates the ignition spark at the plug electrodes. The SCR commutates as the capacitor voltage falls through zero and current from the collapsing field attempts to flow in the reverse direction through the SCR. The cycle then repeats.

The magneto-type CD ignition system shown in Fig. 5-28 is found in chain saws, lawn mowers, snowmobiles, and a variety of other devices powered by small engines. There are no breaker points to wear out, get out of time, get dirty, or adjust. With no moving parts, there is nothing to foul up or mechanically wear out. Everything is encased in a solid-plastic waterproof, dirtproof, vibrationproof module. Practically no servicing is ever required, except for occasional spark plug replacement.

This revolutionary system generates voltage and current, stores power, and creates and times the spark. It is said to give a hotter spark—up to 30,000 volts—for faster starts, better performance, and, according to the manufacturer, to provide more than double normal spark plug life.

CHAPTER 6

Troubleshooting Solid-State Ignition Systems

Replacing a conventional Kettering ignition system with a solid-state ignition system means changing a number of practical adjustments, measurements, and troubleshooting techniques that, previously, could be applied to fixing ignition problems in virtually any car. With the installation of a solid-state system, servicing and troubleshooting become very particular, indeed. Procedures that reveal problems in a transistorized contact-triggered, inductive-discharge system will be considerably different from the steps needed to isolate faults in a capacitive-discharge system. In this section, we will present a series of servicing and troubleshooting tips keyed to the wide variety of system types now available in the family of solid-state ignition.

PROBLEMS COMMON TO MOST SYSTEMS

Fortunately, be it conventional or solid state, virtually any ignition system you come across will have a secondary high-voltage circuit that hardly varies from car to car. Spark plugs, plug wires, distributor cap, center-post wire insulators, rotor, and ignition coil are components common to almost all systems. Understanding the nature of the problems that each can cause is important in accurately diagnosing a failure.

Spark Plugs

Most spark plug "failures" can be attributed to excessive gap growth or fouling deposits, as shown in Fig. 6-1. These conditions are usually traceable to poor fuel system operation or erratic firing. The plug can be the villain or the victim. If it is damaged during installation, it can harbor a leakage path between the tip and the shell, bypassing the spark potential around the firing gap. More likely, however, the plug will simply fire intermittently due to another component that has failed. Unburned or partially burned fuel deposits will coat the plug insulator, eventually bridging the gap and rendering the plug "unfirable." Ignition systems with a fast spark rise time tend to work against this type of failure since the gap potential reaches the arc point faster than shunting deposits can drain away energy. Even so, such a system can mask a defect in the secondary circuit only until insulation resistance breaks down somewhere else—usually in the spark plug cables, ignition coil, or distributor cap.

As a rule of thumb, spark plug wear averages 0.1 millimeter per 1000 miles in a conventional ignition system. Solid-state inductive-

Fig. 6-1. A badly fouled spark plug.

discharge systems have a much longer arc period than capacitive-discharge systems and will produce the greatest spark plug gap erosion. Thus, although it is practical to clean and regap used plugs once in most systems with good results, the probability of getting away with it a *second time* is virtually nil in solid-state inductive-discharge systems.

Spark Plug Cables

Few car owners routinely inspect or replace spark plug cables, and yet these are the principal offenders in "miss" problems. Most of these cables are not wires, but special radio-interference-suppres-

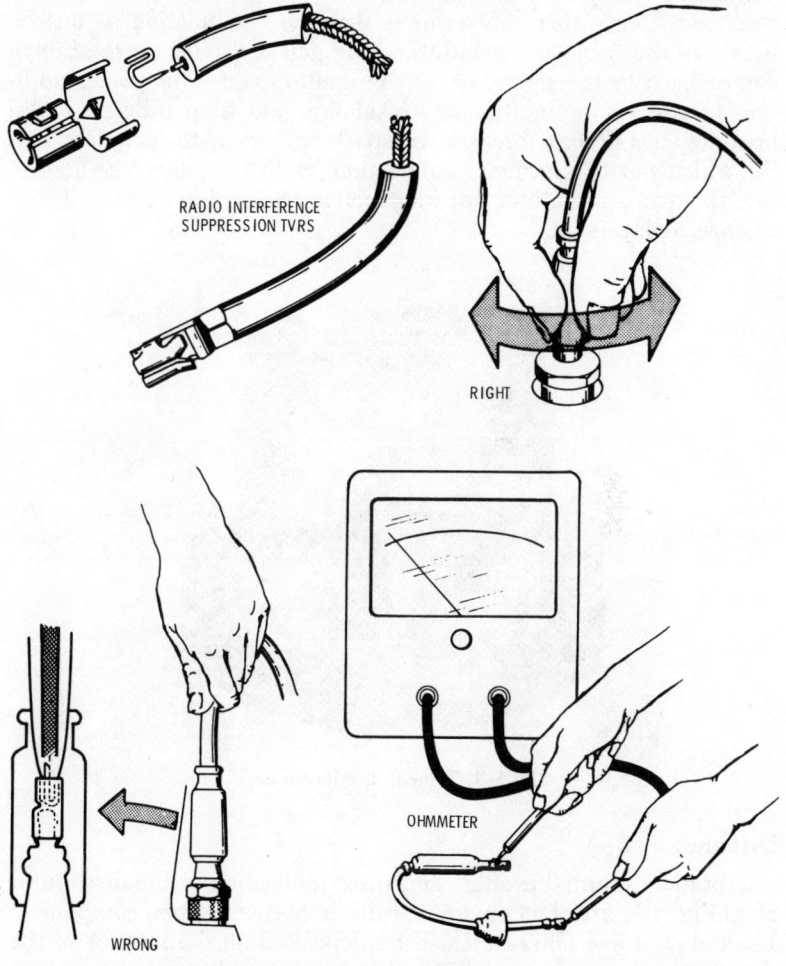

Fig. 6-2. Spark-plug cable removal and testing.

sion (TVRS) cables. These cables usually consist of a nylon or fiberglass core impregnated with a resistive composition material within an insulating jacket. Heat, vibration, and rough handling are very hard on these cables, particularly after they have been in place a year or so and have "taken a set." Roughly removing a plug wire from a plug can cause an open circuit in the brittle inner conductor, as shown in Fig. 6-2. Now, when the wire is reconnected, *two gaps* must be jumped by the high voltage from the ignition system. The attenuation of the spark plug electrode voltage is sufficient to cause erratic firing. Of course, if the insulation of the cable has deteriorated along its length, it is possible that spark voltage will simply jump to the engine block and never fire the plug. The worst possibility that can occur is that the combination of unseen breaks in the cable along with the plug gap will reflect a *total open circuit* back to the secondary of the ignition coil. This places additional stress on the ignition coil insulation and frequently results in breakdown. Often, a breakdown spark will seem to jump *through* the coil tower to a primary connection, or to the case. Needless to say, this cannot continue for long before the ignition coil will also require replacement.

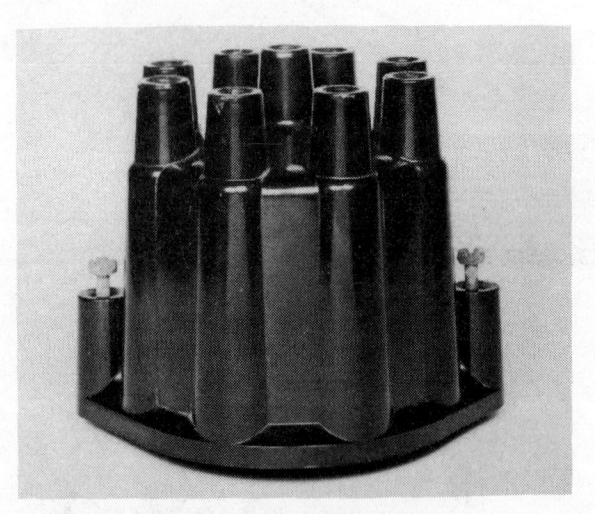

Fig. 6-3. Typical distributor cap.

Distributor Cap

Vibration, thermal cycling, and fume infiltration of the distributor cap (Fig. 6-3) are chief enemies of this important system component. Inevitably, a fine film of oil will be deposited on the interior of the distributor cap and degrade the insulation resistance between one or more of the spark plug wire terminals in the cap and the grounded distributor body. While not overly harmful in itself, this high-resistance pathway can prove troublesome if it offers a better path to ground than, say, the combination of a worn spark plug and aged, broken cable. Successive arc-overs tend to carbonize the path in the cap and reduce its resistance, so that occasional erratic misfiring gradually gives way to an aggravating steady miss.

Thermal cycling and vibration also have a way of shaking down minute faults in the molded construction of the distributor cap. Microscopic voids in the cap establish natural lines of weakness that can physically separate, producing a hairline crack of between two of the insulated terminals or between a terminal and ground. Moisture inevitably seeps into the crack, providing a conduction path. If the path is sufficiently conductive, a spark can be applied to two or more plugs simultaneously. Since one of the two pistons is not yet at firing position, preignition and destructive "knock" can occur. Less dire is the consequence of a crack from a cap terminal to ground, since it produces a misfire similar to that described previously for a spark plug cable with defective insulation.

Rotor

Failure of a distributor rotor is rare, although it can be helped along by improper installation or gross physical abuse. As with the distributor cap, the molded rotor is susceptible to thermal cycling, vibration, moisture, and oil filming.

Coil-to-Distributor High-Tension Lead

This is the lead that connects the ignition coil to the center post of the distributor. Ordinarily made of the same composition material used for the spark plug cables, this key lead can easily open, giving erratic misfire symptoms or a "no-start" condition. Moisture-proofing boots at either end of the lead are subject to dry cracks after prolonged exposure to engine heat. These can lead to moisture entry and eventual corrosion of the coil or cap terminals, with misfiring the inevitable result.

Ignition Coil

Ruggedly constructed and filled with epoxy, pitch, or oil, the modern ignition coil is among the most reliable of automotive components. Ignition-coil failures typically result from faulty manufacture or from an allied failure of a secondary ignition-system component. Few insulating materials retain a constant dielectric strength after being continuously stressed by high voltage. This is true of materials used in ignition coils. An open secondary circuit presents a "no-load" condition to the coil winding, allowing voltages to soar too far in excess of usual plug firing voltages. The coil, which constantly "looks" into a secondary system having one or more open plug wires or excessively gapped spark plugs, is under great stress. Eventually, this stress may be relieved by an internal arc-over between layers in the secondary winding. Once this has occurred, the dielectric strength of the insulating material in the coil is forever lost, and the internal path will provide a better arc path than even a closely gapped external spark plug circuit. Replacement of the ignition coil is then the only alternative.

The arc-through occurring between the ignition-coil tower and a primary connector or ground need not be considered the end of a coil's useful life, *if* the condition is remedied early. Typically, this is a symptom of an open ignition secondary circuit, and proper remedy can prevent deterioration of the coil insulation.

TROUBLESHOOTING THE IGNITION SECONDARY CIRCUIT

The following procedures present typical symptoms, probable causes, and suggestions for isolating faults in high-voltage secondary circuits. Regardless of what type of ignition system is driving the primary circuit, you should first satisfy yourself that one of these common faults is *not* present before delving into the solid-state circuitry on the primary side. It is, of course, also essential to be sure that a fault does not exist in the fuel system or some other mechanical operation before needlessly attempting to troubleshoot the ignition system. No amount of tinkering with the ignition system can compensate for a defect in fuel-air delivery or the loss of compression, on which operation of the internal combustion engine is founded. Common sense, good judgment, the vehicle shop manual, and the procedures presented here should be regarded as helpful resources in locating common faults that *do not* originate in the solid-state portion of an ignition system.

Engine Cranks But Will Not Start

The probable causes for this symptom include a defective or grounded coil-to-distributor lead, a defective ignition coil, a defective rotor, a defective distributor cap, or fouled or excessively worn spark plugs. Use the following procedures to determine the source of the problem.

Procedure 1. Remove the ignition-coil lead from the center post of the distributor and place it next to the engine block. As the engine is cranked, a spark should jump to the block. If there is no spark, measure the resistance of the lead. Replace the lead if it is open. Typically, the normal resistance of this lead is approximately 10,000 ohms. Visually examine the lead to be sure it has not been worn by abrasion or is not grounded in any way.

Procedure 2. As the engine is cranked, check the ignition coil in a dark area for external arc-over. If there is no indication of arc-over, measure the resistance of the coil secondary winding. Replace the coil if it is open or if zero resistance is indicated between the tower terminal and ground (case). Check for internal breakdown by substituting a known-good coil in place of the suspected one. If the system functions normally, replace the original coil.

Procedure 3. Remove the rotor and examine it under a strong light for hairline cracks, signs of arcing, or deformation of the leaf contact. Replace the rotor if any defect is found. (The leaf contact may be realigned manually to the internal center-post wiper of the distributor cap, if necessary.)

Procedure 4. Remove the distributor cap, carefully tagging the spark plug leads for proper reconnection later. Examine the cap under a strong light for cracks or an arc path. Check for a missing inner wiper contact on the center-post terminal of the cap, or for foreign matter that might shunt the high voltage around the rotor. Clean or replace the rotor if necessary.

Procedure 5. Remove the spark plugs and check the condition of the insulator and the electrodes. Clean away any fouling or bridging deposits that may be present. Check the spark plug gap and file or reset as necessary. If one or more of the spark plugs show excessive wear, replace the entire set.

Engine Runs But Misses at Idle

The most likely causes of this symptom are open or grounded spark plug leads, a fouled or defective spark plug, or a defective distributor cap. Use the following procedures to locate the defective ignition component.

Procedure 1. With the engine running at idle, carefully remove one spark plug lead at a time and note the effect upon the smoothness of the engine. Be suspicious of any lead which seems to show *no* difference whether it is connected or not. Measure the lead resistance with an ohmmeter. Replace the lead if it is open. Also check for a lead that is grounding to the block or accessories. Replace or redress the lead as necessary. (Consider replacing the entire spark plug cable harness if the leads are brittle or generally deteriorated.)

Procedure 2. Check for a defective spark plug, using the same method as outlined for locating a defective spark plug cable in Procedure 1. Remove and inspect the suspected plug. If the plug is not obviously defective, check it for fouling or bridging deposits. If there are no indications of fouling, check electrodes for edge squareness and gap width. If necessary, file and reset the gap. (Presence of fouling deposits may also signify mechanical deterioration of engine parts, especially in older engines. Suspect this source if no other defect can be found in the secondary or primary of the ignition system.)

Procedure 3. Refer to Procedure 4 under *Engine Cranks But Will Not Start*, for checking distributor cap.

Engine Runs Fair With Occasional Missing

The causes for this symptom include defective spark plug or ignition-coil cables, defective distributor cap, defective ignition coil, fouled or worn spark plugs, and fuel system failure. Use the following procedures for locating the defect.

Procedure 1. Refer to Procedure 1 under *Engine Cranks But Will Not Start*, for checking spark plug and ignition-coil cables.

Procedure 2. Refer to Procedure 4 under *Engine Cranks But Will Not Start*, for checking the distributor cap.

Procedure 3. Refer to Procedure 2 under *Engine Cranks But Will Not Start*, for checking the ignition coil.

Procedure 4. Refer to Procedure 5 under *Engine Cranks But Will Not Start*, for checking the spark plugs. Also see Procedure 2 under *Engine Runs But Misses at Idle*.

Procedure 5. Check the fuel system. Investigation of the fuel system exceeds the scope of this book. Refer to the vehicle shop manual for the proper procedure.

TROUBLESHOOTING THE IGNITION PRIMARY CIRCUIT

Once you have eliminated the common ignition secondary circuit and you begin examining the systems driving the primary of the ignition coil, very real differences in fault-analysis and troubleshooting procedures become evident. A logical procedure is to first *iden-*

tify the system by type (see Chapter 4). The remainder of this chapter is divided into procedures useful in troubleshooting inductive-discharge and capacitive-discharge systems, and troubleshooting information is presented in the same order in which these systems were described in Chapters 4 and 5. Familiarize yourself with the normal operation of a particular type of system beforehand to aid your efforts to determine what has gone wrong.

CONTACT-TRIGGERED INDUCTIVE-DISCHARGE SYSTEMS

If you think of these systems as "transistorized" conventional systems, you will be able to apply much of what you already know about ignition-system troubleshooting to resolving problems that may occur. Almost all such systems rely upon the ignition breaker points for *spark timing*, and many rely on the points for *dwell* functions as well. Thus, the consequences of mechanical wear of the breaker points or rubbing block in such a system relate directly to what happens in a conventional system. Armed with this information, you can attack a faulty contact-triggered inductive-discharge system and localize the fault in minutes using a volt-ohm-milliammeter (vom) and some quick visual checks. Assuming that you have eliminated the ignition secondary circuit as the source of the problem, check the primary circuit as follows.

Checking Ignition-Coil Primary Voltage

This test quickly determines whether or not sufficient voltage is being delivered to the primary of the ignition coil by the solid-state ignition system while the engine is cranking. A normal reading confirms that the general condition of the battery, the cables, the starting system, and the circuitry of the ignition system is satisfactory. An unsatisfactory reading indicates that further testing is required. To perform the test:

1. Connect the leads (+) and (−) of the vom (set to volts) to the ignition-coil primary terminal (+) and ground.
2. With the ignition switch on, crank the engine for five seconds and observe the meter reading. The voltage should not be less than 7.5 volts (average).

If the meter reads the specified voltage or more and the cranking speed is normal and even, the battery, starter, cables, switch, and solid-state ignition circuit to the coil are operating satisfactorily.

If the meter reads less than specified or zero voltage, suspect a weak battery; a defective ballast, resistor cable, connection, switch, starter, or bypass "condenser"; a faulty breaker-point circuit; or defective ignition-system circuitry.

A zero-volt reading at the primary terminal of the ignition coil *may* mean trouble in the transistor switching circuitry of the electronic ignition unit. However, do not overlook the simple possibility that the 12-volt supply line may have grounded or opened. The slip-on, quick-disconnect terminals now widely used in engine compartment wiring can inadvertently disconnect themselves if the wiring is moved about roughly. With the battery disconnected, run ohmmeter checks between the 12-volt supply line to the ignition electronics and ground. Resistance readings of at least several thousand ohms should be noted. If not, trace the fault by visually inspecting the supply line. Similarly, check the supply line continuity back to the 12-volt source with the ignition switch closed. Also, check the ballast resistor which should have a typical resistance of 0.5 to 2 ohms. Be sure to check the primary of the ignition coil for continuity. Reconnect the battery cable after tests and/or repairs are made.

Checking Breaker Points and Electronic Ignition Unit

If you have satisfied yourself that the fault is related to the electronics, make this simple check:

1. Move the vom (+) lead to the (−) terminal of the ignition-coil primary; connect the vom (−) lead to ground.
2. Gain access to the distributor interior. "Spot" the engine, if necessary, to close the breaker points. Leave the ignition switch on.
3. With the breaker points closed, observe that the meter indicates approximately the vehicle supply voltage. Now, manually open the points. Observe that the meter reading falls to and remains at zero. Repeat this test several times. (NOTE: in a dwell-extender-type system, the fall to zero is immediately followed by a return to the previous supply level, regardless of whether the points are reclosed or remain open. The few milliseconds of downward deflection are so brief that your meter may average the drop. In such a system, look for a slight downscale deflection each time the points are opened.)

If the meter reading drops to zero (contact-triggered systems) and returns to supply voltage as points are cycled from closed to open, the electronic ignition unit is good. If the meter reading deflects down-scale slightly (dwell-extender systems) and immediately returns to the supply-voltage level each time the points are opened, the dwell-extender electronic ignition unit is good.

If the meter reading does not change as the points are cycled (open/close), look for defective wiring between the points and the electronic ignition unit, shorted points, faulty bypass "condenser" shunting points, or a defective electronic ignition unit.

Before removing the electronic ignition unit, be certain that there are no short-circuit paths bridging the points, nor something as simple as a defective lead between the points and the electronic ignition unit. Once a fault has been isolated to the electronics, further checks must be made according to the manufacturer's service literature or by using the schematic for a particular design. The circuits provided in Chapters 4 and 5 will be helpful in troubleshooting specific units.

System Runs But Performance Is Poor

The dependence of contact-actuated systems on the condition of the breaker points makes this component a logical source of most poor-performance complaints. Points *do* wear in these applications and should be visually inspected for pitting, "bluing," or mechanical wear. A worn rubbing block means a shift in timing and dwell in many systems. Renewing the breaker-point set is thus an effective cure. In setting up a new point set for the correct gap, a dwell-tachometer such as that shown in Fig. 6-4 can be used. In most

Fig. 6-4. Accurate Instrument Model BT-162 dwell-tachometer.

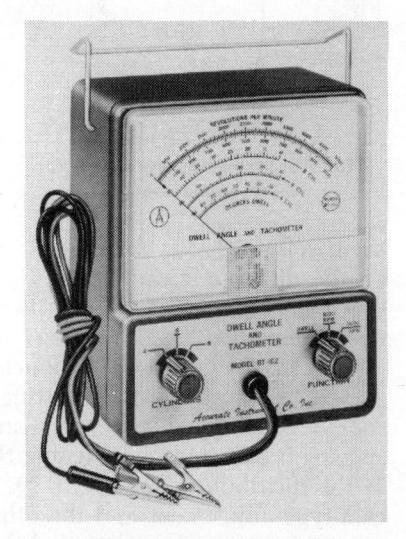

Courtesy Accurate Instrument Co., Inc.

cases, it is recommended that the ignition system be restored to conventional Kettering operation while the points are being set up. Most electronic ignition units provide a change-over switch for just that purpose. However, if your system lacks such a switch and temporary rewiring is too great a chore, you may have to resort to me-

113

chanical gapping of the points by a feeler gauge as the only practical method for setting the dwell. Connections for dwell adjustment with the electronic ignition unit switched out of the circuit are shown in Fig. 6-5.

With the engine idling and the ignition system restored to conventional operation, direct dwell readings may be made by use of the appropriate meter scales. Typical settings are 30° on 8-cylinder engines, 40° on 6-cylinder engines, and 60° on 4-cylinder engines. Always consult the manufacturer's service manual for the proper dwell settings. If the dwell reading is not within ±4° of the angle specified, a correction needs to be made in the point setting. On most General Motors cars, breaker-point dwell setting may be adjusted as shown in Fig. 6-6 by the use of an Allen wrench. To adjust

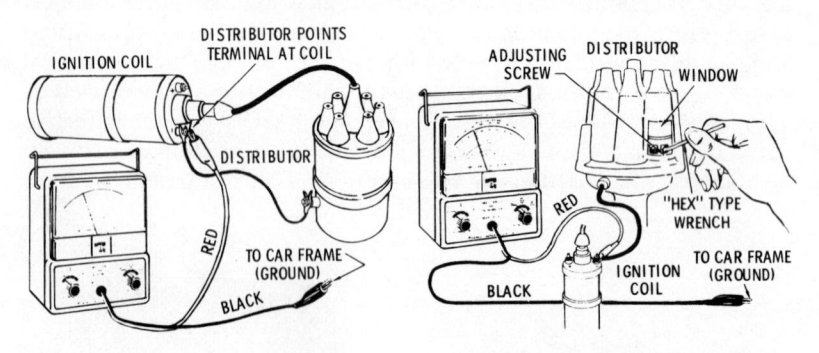

Fig. 6-5. Standard connections for dwell measurement.

Fig. 6-6. Dwell adjustment of General Motors type distributors.

the dwell on these vehicles, it is necessary to lift the metal window on the distributor cap. Insert the Allen wrench into the socket on the point set and adjust the dwell by slowly rotating the wrench until the correct dwell readings are obtained on the meter.

On Chrysler and Ford products, the distributor cap must be removed to adjust the dwell setting. To adjust the points, the engine should be stopped and the distributor cap lifted off. The point spacing is adjusted by loosening the two screws that hold the points in the distributor (see Fig. 6-7). Generally, one screw serves as a cam-type adjustment and the other serves as a locking screw. The points should be loosened slightly and a careful adjustment made. To decrease dwell, the point gap must be increased; and to increase dwell, the gap is decreased. If a feeler gauge is available, typical breaker-point gap is 0.015 inch for proper dwell. The points are then tightened and the distributor cap replaced. The dwell should be checked again by starting the engine and using the dwell meter as explained before. After correction adjustments are made in the

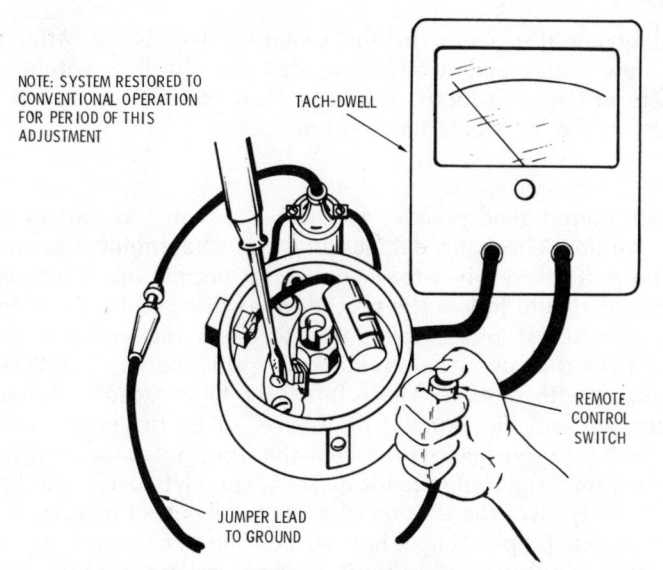

Fig. 6-7. Proper method for setting dwell on a distributor with internal adjustment.

dwell setting, the engine timing should be checked and any necessary corrections made in order to insure proper timing and dwell. If the dwell readings vary more than 3° from the initial reading between idle speed and approximately 1500 rpm, the distributor shaft is probably worn. However, variations at speeds above 1000 rpm may not be wear-induced, but may be normal to specific engine designs. Some engine manufacturers intentionally change the dwell angle at approximately 1000 rpm. If the engine under test exhibits a significant change in dwell angle, it is best to check the manufacturer's tune-up specifications to determine whether the change is a design feature or a fault.

On present-day engines, it is very important to set basic timing exactly to specifications. A few years ago, most engines would tolerate a moderate amount of deviation from specifications to compensate for fuel octane and other variables. However, present emission-controlled models simply will not tolerate incautious tinkering with the timing. Because the timing is affected by the dwell, as well as by the secondary voltage, it should be rechecked whenever the point gap is adjusted. If the breaker-point gap is too wide, the points open sooner, so ignition timing is actually advanced. This reduces the dwell angle which may decrease the secondary voltage, causing a miss at higher engine speeds, even with electronic ignition. On the other hand, a narrow point gap increases dwell. This may result in rough engine operation at low speeds, with attendant poor plug

115

life and higher dissipation by the switching transistors. After installing new points and determining that the dwell is within the manufacturer's specifications, you may then restore the system to operation on the electronic ignition unit.

Timing

It is considered good practice to check the timing as part of any general ignition check or engine tune-up. The timing operation should be performed when the engine is at normal operating temperature and should follow the procedure as directed by the manufacturer. The surest and simplest way to check the timing is with a timing light that uses the stroboscopic principle of light flashes synchronized with motion. These bursts of light are of extremely short duration and they appear to "freeze" or fix the position of a moving pulley or flywheel with which the light pulses are synchronized. Each time the timing mark makes a complete cycle, the light flashes on. This gives the illusion of arresting flywheel motion, even while the engine is operating at high speed. The timing mark appears to be standing still, which makes it a simple matter to observe the effects of ignition timing adjustments. If the mark and the reference point do not align, the hold-down screw for the distributor must be loosened and the distributor rotated, either counterclockwise or clockwise, until proper timing-mark alignment is obtained.

The location of the reference pointer or timing mark can usually be determined from the manufacturer's specifications. It may be in any one of several places such as on the harmonic balancer, vibration damper, flywheel, crankshaft, impulse neutralizer (located on the front of the engine), or it may be a line of steel balls embedded in the flywheel at the rear of the engine. Typical timing marks are shown in Fig. 6-8. The timing mark can be located by cranking the engine slowly until it appears. For increased visibility, it is ad-

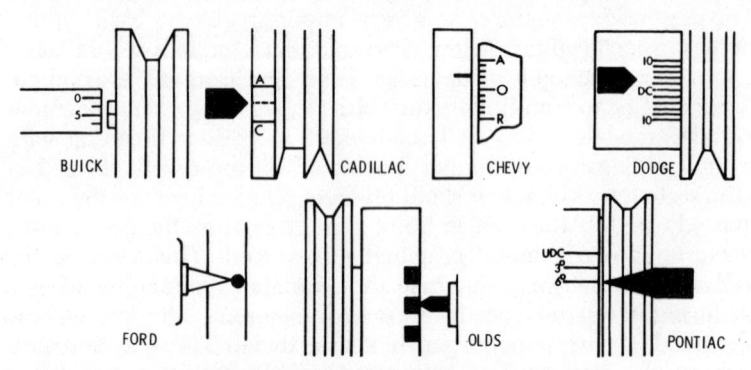

Fig. 6-8. Types of timing marks in common use.

Courtesy Radio Shack

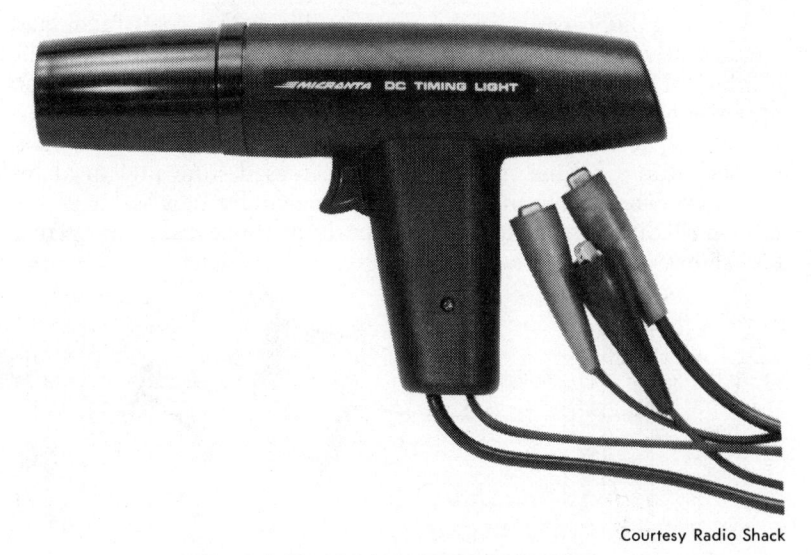

Fig. 6-9. Typical dc-powered timing light.

visable to make a white line about ⅛ inch wide directly over the timing mark. This helps locate the mark when the engine is running during a timing check.

A xenon-lamp dc powered timing light of the type illustrated in Fig. 6-9 produces a high intensity, short-duration flash and makes the timing adjustment easy. This type of timing light contains a source of high voltage that charges an energy-storage capacitor, which then discharges into the xenon lamp every time it is triggered by the ignition pulse at the spark plug.

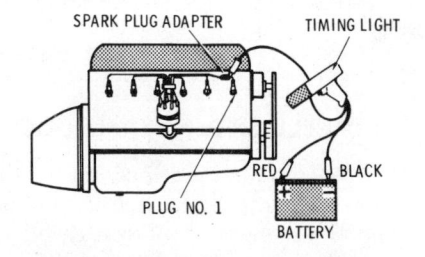

Fig. 6-10. Dc-powered timing-light connections.

In performing the basic timing check, the timing light is aimed at the timing marks while the distributor is manually rotated. When the illuminated marks achieve proper alignment (with engine running at specified rpm), the distributor hold-down bolt is tightened. Fig. 6-10 illustrates the connection and proper use of timing light.

117

A timing light can also be used to check the centrifugal and vacuum advance. As the engine speed is slowly increased from idle to about 1500 rpm, the timing mark should move in the direction opposite to engine rotation, as shown in Fig. 6-11. Note how much movement occurs. The vacuum line to the distributor is then disconnected and engine speed is again increased from idle to about 1500 rpm. The timing mark should still move, but not as far as the first time. Since the vacuum advance is disconnected, this second test shows centrifugal advance only.

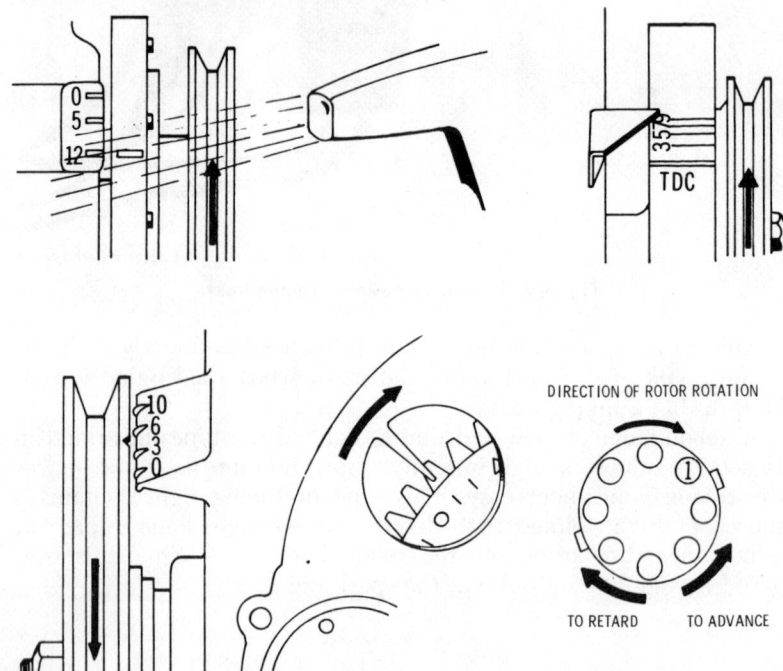

Fig. 6-11. Using the timing light.

DWELL-EXTENDER INDUCTIVE-DISCHARGE SYSTEMS

The systems described under this heading in Chapter 4 ranged from a simple SCR "helper" to the antiarcing *Electronic Magneto* and the elaborate *Solitron* dwell extender. Of these, the SCR-type system is the easiest to troubleshoot.

SCR Dwell Extender

Disconnect the dwell-extender connecting lead from the screw post on the breaker-point set. This restores the ignition system to conventional operation. If the engine will start and run, the fault lies in the dwell-extender unit and can be isolated by resistance checks with a vom. A shorted SCR is a typical failure in these systems.

Electronic Magneto System

Perform the tests described under the headings *Checking the Ignition-Coil Primary Voltage* and *Checking the Breaker Points and Electronic Ignition Unit* for the contact-triggered inductive-discharge systems. If the wiring and breaker points check good, troubleshoot the electronic unit. Failure of the transistor or either of the zener diodes will render the unit inoperative. The continuity of the coil can be checked between the accessible terminals on the unit.

Solitron Dwell Extender

Perform the same initial test as described for the Electronic Magneto system. Remember that the "off" time for this circuit is very brief. Try operating the system with switch S1 set for conventional ignition. If satisfactory operation is obtained, it will be necessary to troubleshoot the electronic unit. (NOTE: The electronic unit is epoxy filled, which makes disassembly unusually difficult.)

MAGNETICALLY TRIGGERED INDUCTIVE-DISCHARGE SYSTEMS

Magnetically triggered systems are the universal choice of major manufacturers and a few manufacturers of aftermarket systems. Most typical of the current designs are those by Chrysler and Ford.

Chrysler Electronic Ignition

When troubleshooting this system, the first step is to make a mechanical inspection of the system components. Failure in the distributor unit is rare, but the magnetic pickup can be physically misadjusted. Check the air gap (Fig. 6-12) between the reluctor tooth and the pickup coil. Loosen the pickup hold-down screw and insert a .008" nonmagnetic feeler gauge between the reluctor tooth and the pickup coil. Adjust the pickup so that the .008" feeler gauge is snug and then tighten the hold-down screw.

Visually inspect all secondary cables at the ignition coil, distributor, and spark plugs for cracks and tightness. Check the primary wires at the ignition coil and ballast resistor for tight connections. If the above checks do not resolve the problem, the following electrical checks will determine if a component is faulty.

1. Unplug the wiring-harness connector from the control unit. Fig. 6-13 shows the connector cavities (female pins). Turn the

HOLD DOWN SCREW

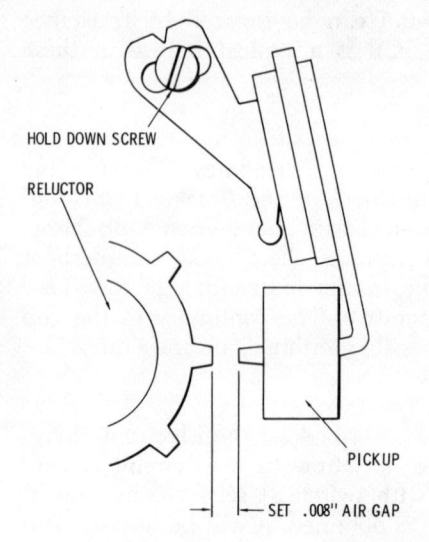

RELUCTOR

PICKUP

SET .008" AIR GAP

Fig. 6-12. Mechanical adjustment of air gap in Chrysler distributor.

ignition switch on. Connect the negative lead of the voltmeter to a good ground.
2. Connect the positive lead of the voltmeter to harness connector cavity No. 1 (see Fig. 6-13). The voltage at cavity No. 1 should be within one volt of the battery voltage with all the accessories turned off. If the voltage is less, check the circuit back through the ballast resistor and ignition switch to the battery, as shown in Fig. 6-14.
3. Connect the positive lead of the voltmeter to the wiring-harness connector cavity No. 2. The voltage at cavity No. 2 should be within one volt of the battery voltage with all the accessories turned off. If there is more than a one volt difference, Fig. 6-14 shows the circuit that must be checked.

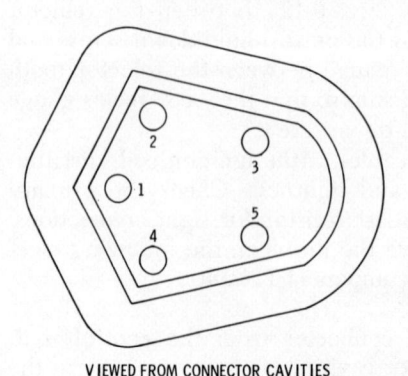

Fig. 6-13. Wiring-harness connector cavities.

VIEWED FROM CONNECTOR CAVITIES

4. Connect the positive lead of the voltmeter to the wiring-harness connector cavity No. 3. The available voltage at cavity No. 3 should be within one volt of the battery voltage with all the accessories turned off. If there is more than one-volt difference, Fig. 6-14 shows the circuit that must be checked. Turn the ignition switch off.

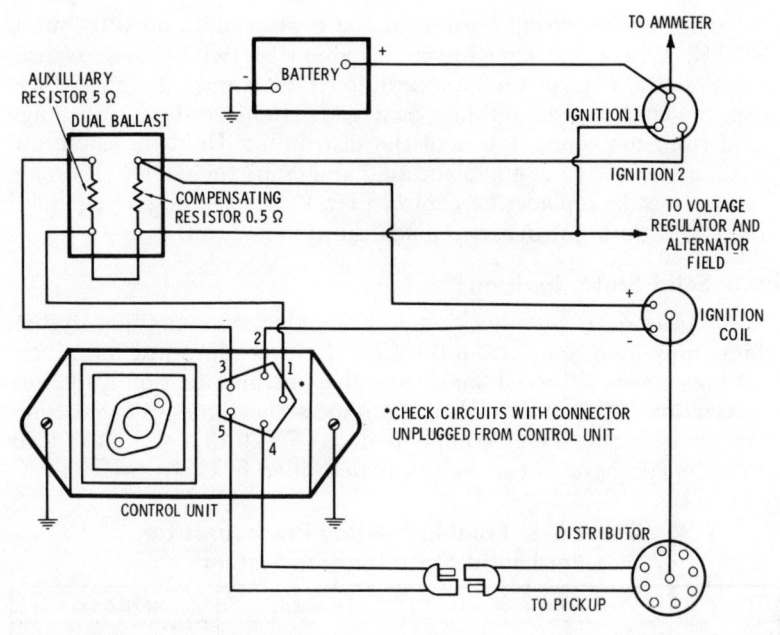

Fig. 6-14. Wiring diagram of Chrysler electronic ignition system showing paths to check for defects.

5. Connect an ohmmeter between wiring-harness connector cavities No. 4 and No. 5 (Fig. 6-13). The ohmmeter reading should be between 350 to 550 ohms. If the reading is higher or lower than specified, disconnect the dual-lead connector coming from the distributor and check the resistance back through the distributor. If the reading is not between 350 and 550 ohms, replace the pickup-coil assembly in the distributor. If the reading is within specifications, check the wiring harness from the dual-lead connector back to the control unit. Connect one ohmmeter lead to a good ground and the other lead to either pin of the dual-lead connector from the distributor. The ohmmeter should indicate an open circuit. If the ohmmeter shows continuity, the wiring or the pickup coil in the distributor is grounded. Repair or replace, as necessary.

121

6. Connect one ohmmeter lead to a good ground and the other lead to control-unit connector pin No. 5 (Fig. 8-14). The ohmmeter should show continuity between ground and the connector pin. If continuity does not exist, tighten the bolts holding the control unit to the firewall, then recheck. If continuity still does not exist, the control unit must be replaced.

Reconnect the wiring harness at the control unit and distributor. (NOTE: Whenever removing or installing the wiring-harness connector at the control unit, the ignition switch must be off. Otherwise, the control unit could be damaged.) Remove the high-voltage cable from the center tower of the distributor. Hold the cable approximately $\frac{3}{16}''$ from engine ground and crank the engine. If arcing does not occur, replace the control unit. Crank the engine again. If arcing still does not occur, replace the ignition coil.

Ford Solid-State Ignition

The procedure for checking a Ford electronic ignition system differs only in specifics from the Chrysler procedure just described.

If no spark is observed, make sure that the high-tension lead from the ignition coil is good. Then disconnect the three-way and four-way connectors at the electronic module (Fig. 6-15), and make tests at the female harness connectors as described in Chart 6-1.

Chart 6-1. Troubleshooting Procedures for Ford Solid-State Ignition System

	Test Voltage Between	Should Be	If Not, See
KEY ON CRANK- ING	Pin 3 and Engine Ground Pin 5 and Engine Ground Pin 1 and Engine Ground Pin 5 and Engine Ground Pin 7 and pin 8	Battery Voltage Battery Voltage 8 to 12 volts 8 to 12 volts ½ volt ac or dc	Module Bias Test Battery Source Test Cranking Test Starting Circuit Test Distributor Hardware Test
	Test Resistance Between	Should Be	If Not, See
KEY OFF	Pin 7 and Pin 8 Pin 6 and Engine Ground Pin 7 and Engine Ground Pin 8 and Engine Ground Pin 3 and Coil Tower Pin 5 and Pin 4 Pin 5 and Engine Ground Pin 3 and Pin 4	400 to 800 ohms 0 ohms more than 70,000 ohms more than 70,000 ohms 7000 to 13,000 ohms 1.0 to 2.0 ohms more than 10.0 ohms 1.0 to 2.0 ohms	Magnetic Pickup (Stator) Test Coil Test Short Test Resistance Wire Test

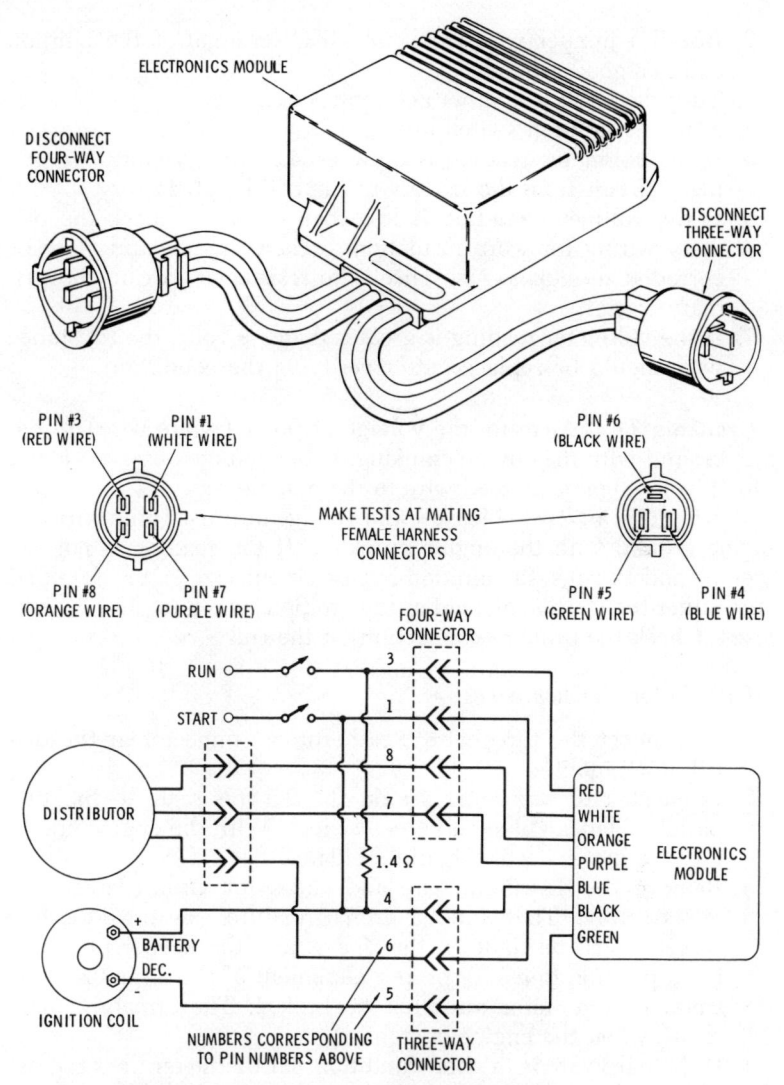

Fig. 6-15. Wiring diagram and circuit identification for checking Ford solid-state ignition system.

Module Bias Test—Measure the voltage at pin 3 (red wire) to engine ground with the ignition key on. If the voltage observed is less than the battery voltage, repair the feed wire to the module.

Battery Source Test—
1. Connect the voltmeter leads from the battery terminal at the ignition coil to engine ground without disconnecting the coil.

123

2. Install a jumper wire from the DEC terminal of the ignition coil to a good ground.
3. Turn the lights and other accessories off.
4. Turn the ignition switch on.
5. If the voltmeter reading is between 4.9 and 7.9 volts, the primary circuit from the battery to the coil is satisfactory.
6. If the voltmeter reading is less than 4.9 volts, check the primary wiring for worn insulation, broken strands, and loose or corroded terminals. Also check the resistance wire in the primary circuit.
7. If the voltmeter reading is greater than 7.9 volts, the resistance wire should be replaced after verifying the condition.

Cranking Test—Measure the voltage at pin 1 (white wire) to engine ground with the engine cranking. If the voltage observed is not 8 to 12 volts, repair the feed wire to the module.

Starting Circuit Test—Measure the voltage at pin 4 (blue wire) to engine ground with the engine cranking. If the reading is not between 8 and 12 volts, the ignition bypass circuit is open or grounded from either the starter solenoid or the ignition switch to pin 5 (green wire). Check the primary connections at the coil.

Distributor Hardware Test—

1. Disconnect the three-wire weatherproof connector at the distributor pigtail.
2. Connect a dc voltmeter set on the 2.5-volt scale to the two parallel blades (blue and green wires). With the engine cranking, the meter needle should oscillate.
3. Remove the distributor cap and check for visual damage or misassembly. The sintered iron armature (6- or 8-toothed wheel) must be tight on the sleeve, and the keeper pin must be in position to assure proper alignment of the armature. The sintered iron stator must not be broken. The armature must rotate when the engine is cranked.
4. If the hardware is in good condition but the meter does not oscillate, check the magnetic pickup (stator assembly) as described below.

Magnetic Pickup Test—

1. Disconnect the three-wire weatherproof connector at the distributor pigtail.
2. The resistance of the pickup coil measured between the two parallel blades (green wire and blue wire) in the distributor connector must be 400-800 ohms.

3. The resistance between the third blade (black ground wire) and the distributor bowl must be zero ohms.
4. The resistance between either parallel blade and the engine ground must be greater than 70,000 ohms.
5. If any of these tests fail, the distributor magnetic pickup (stator assembly) is inoperative and must be replaced.
6. If the above tests check good, the magnetic pickup (stator assembly) in the distributor is all right.

Ignition Coil Test—The ignition coil must be diagnosed separately from the rest of the ignition system. Use the following tests to check the coil.

1. The primary resistance of the coil must be 1.0 to 2.0 ohms.
2. The secondary resistance must be 7000 to 13,000 ohms.
3. If the coil is still suspected, test it on a coil tester by following the manufacturer's instructions for testing standard ignition coils, or try a substitute coil.

Short Test—If the resistance from pin 5 (*red* wire) to engine ground is less than 10 ohms, check for a short to ground at the DEC terminal of the ignition coil or in the connecting lead to that terminal.

Primary Resistance Wire Test—Replace the primary resistance wire if it is open or of different resistance value than shown in Fig. 6-15.

General Motors/Delco Unitized Ignition

Though the unitized High-Energy Ignition System by Delco-Remy departs radically in appearance from the previously described magnetically triggered systems of Chrysler and Ford, a checkout of the unit on the vehicle follows essentially the standard procedure described earlier. For convenience, a full checkout procedure is presented in Chart 6-2. Before starting this procedure, make sure that the black and pink leads are connected as shown in Fig. 6-16. Also, tighten both bolts (Fig. 6-16); loose bolts may cause radio interference.

CAPACITIVE DISCHARGE SYSTEMS

The nearly instantaneous rise time of CD systems places considerable stress on the high-voltage secondary circuit insulation. Engine miss on acceleration is a good indication that a spark is probably occurring *outside* the engine, where it does no good at all. The checks described earlier in this chapter under the heading PROB-

LEMS COMMON TO MOST SYSTEMS will help reveal a secondary circuit defect.

The primary circuit of the capacitive-discharge ignition system is considerably more hazardous to check than that of the inductive-discharge systems, simply because the energy pulse delivered to the primary of the coil begins at a whopping 400 volts dc, or thereabouts. You should have respect for that potential, for it can be lethal. Never lean across the car in such a way that your chest or

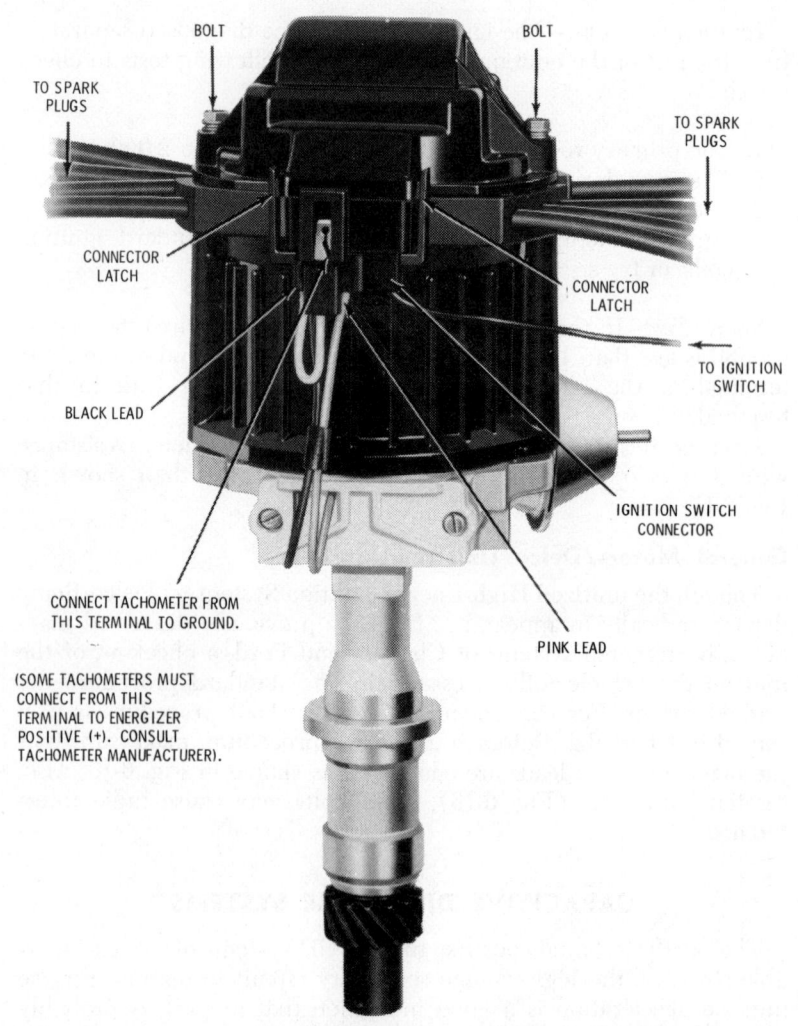

Fig. 6-16. Delco-Remy unitized ignition unit.

Chart 6-2. Troubleshooting Procedure for Delco-Remy Unitized Ignition System

Symptom	Test Procedure
ENGINE WILL NOT RUN AT ALL	1. Check the ignition switch. 2. Connect a voltmeter from the ignition-switch connector to ground. 3. Turn on the ignition switch. 4. If the voltage reading is zero, the circuit is open between the connector and the ignition switch. Repair as needed. 5. If you read the battery voltage at the switch connector, hold one of the spark plug leads about ¼ inch from a dry area of the engine block while cranking the engine. 6. If sparking occurs, the trouble is most likely not in the ignition system. Check the fuel system and the spark plugs. 7. If sparking does not occur, follow the TEST-BENCH PROCEDURE outlined below. Most of this procedure may be performed without removing the unit from the vehicle.
ENGINE STARTS BUT WILL NOT RUN; ENGINE MISSES OR SURGES	1. Make sure that the fuel system is satisfactory. 2. Check the spark plug leads for breaks, arc-over, or leakage to ground. 3. If no defects are found, follow the TEST-BENCH PROCEDURE below. Most of this procedure may be performed without removing the unit from the vehicle.
TEST-BENCH PROCEDURE	1. Disassemble the ignition unit and inspect coil, eight inserts, shell, and rotor for arc-over or leakage. 2. Connect the ohmmeter as shown in Fig. 6-17, using the high scale. 3. The readings for connections A and B should each be practically zero. If either reading is infinite, replace the coil. 4. Connection C should read 6000 to 9000 ohms. If the reading is outside this range, replace the coil. 5. Connection D should give an infinite reading. If not, replace the coil. 6. If the above tests are good, connect a vacuum source to the vacuum unit. Make the ohmmeter connections A and B as shown in Fig. 6-18. Use the middle scale. Observe the ohmmeter readings throughout the vacuum range. 7. If the connection A reads less than 650 ohms or more than 850 ohms at any time, replace the pickup coil as outlined in Step 10 below. 8. If the reading for connection B is other than infinite at any time, replace the pickup coil (Step 10). 9. If the vacuum unit is inoperative, replace it.

Chart 6-2—Continued

Symptom	Test Procedure
	10. To replace the pickup coil, remove the ignition unit from the engine. Remove the drive pin from the gear and the rotor and shaft assembly from the housing. Remove the shim and thin "C" washer to replace the pickup coil (Fig. 6-19).
	11. If no defects have been found, remove the two attaching screws and replace the electronic module.
Note: The hybrid electronic module cannot be repaired and must be replaced if it is defective.	

abdomen is directly grounded to the chassis while you are touching the primary terminals of the CD-fired ignition coil and the engine is running. Under certain circumstances, a shock delivered across this path can disrupt normal heart rhythm and lead to sudden death. More likely is the possibility of injury from involuntary muscular contraction. Play it safe and keep yourself "out of the circuit" by leaning on an insulating blanket and using insulated probes to investigate the primary circuit of a CD-powered ignition system.

In many inverter-type systems, operation of the switching oscillator sets up a sympathetic vibration of the transformer core laminations, generating an audible "whistle" or "squeal" that can be heard by standing near the unit and listening in a quiet location, with the ignition on but the engine not running. However, absence of this acoustical phenomenon is not an infallible indicator that a unit is faulty. Some manufacturers take great pains to suppress this magnetostriction in the inverter circuitry of their products, since it does represent an energy loss. Generally, most units "squeal" but some do not. Chances are, if you hear the inverter running, the fault lies elsewhere.

A neon-lamp circuit of the type shown in Fig. 6-20 makes a handy test aid to determine the presence or absence of ignition-coil primary voltage from the inverter in a CD system. The neon lamp will glow if the potential is greater than 70 volts and will flash brilliantly if the system is then triggered into discharge. The lamp draws a very small current; so, it can read the voltage developed across the capacitor within the CD unit through the high-value bleeder resistor included in virtually every design.

Use the following procedure to check a CD ignition system.

1. If the system is contact-triggered, temporarily short the breaker points by means of a jumper wire from the (−) coil terminal to engine ground. (Skip this step if a magnetically triggered system is being checked.)

Courtesy Delco-Remy Division, General Motors Corporation

Fig. 6-17. Ohmmeter checks of coil.

2. Connect the neon-test-lamp clips between the (+) and (−) terminals of the ignition coil.
3. Turn the ignition switch on, but do not crank the engine.
4. With the voltmeter, check for the presence of battery potential at the unit input leads to the CD unit. If voltage is absent, check for a short or open circuit, or blown fuse.
5. Listen near the CD unit for a telltale "squeal" indicating the

Fig. 6-18. Ohmmeter checks of pickup coil.

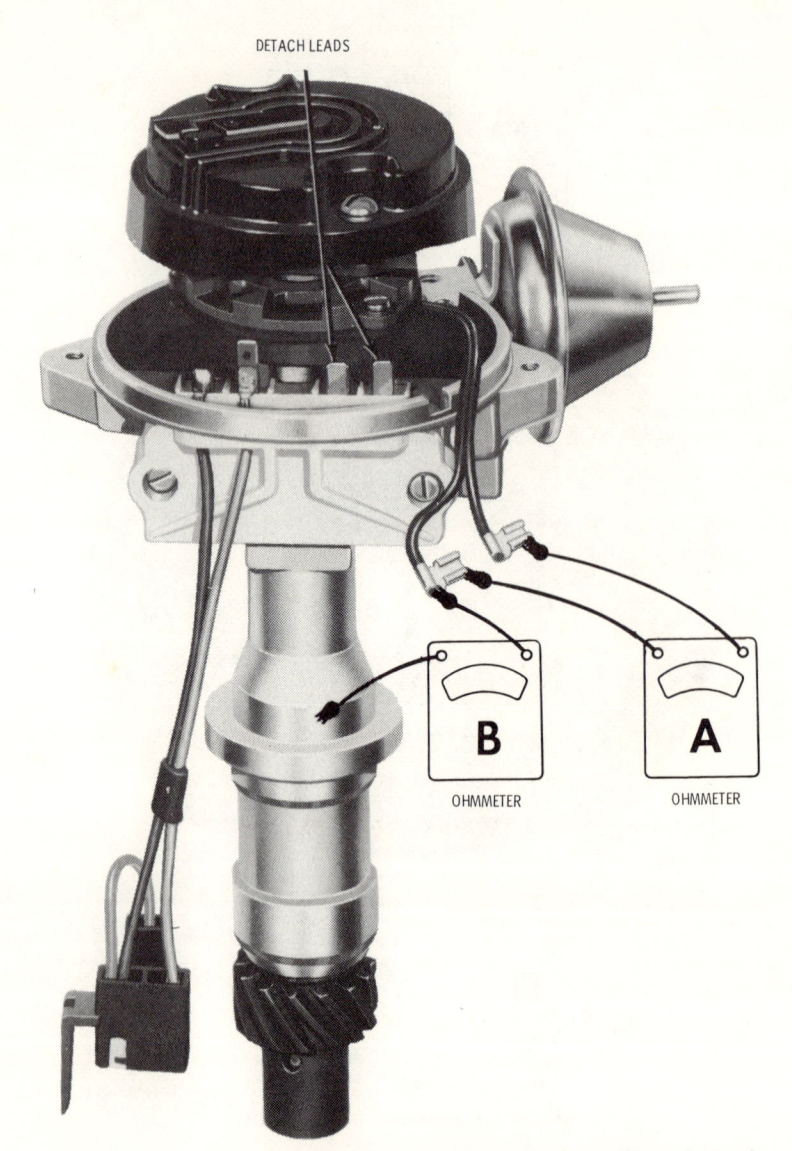

Courtesy Delco-Remy Division, General Motors Corporation

inverter is operating. If you hear nothing, observe the neon test lamp. A glow indicates that the inverter is operating. If neither sound nor glow is detected, an internal fuse may have opened within the CD unit.

6. Block the breaker points open with a nonconductive material.

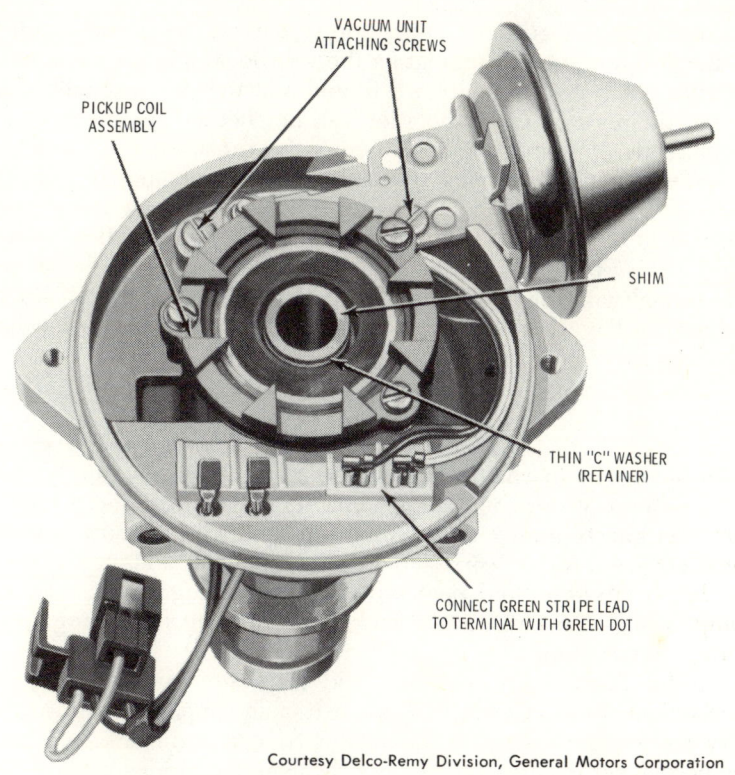

Fig. 6-19. Ignition unit with shaft assembly removed.

While observing the neon test lamp, remove the jumper wire installed in Step 1. The lamp should flash brilliantly, indicating triggering of the SCR and discharge of the capacitor. If the unit fails to trigger, the SCR, capacitor, or trigger circuitry may be defective. Remove the CD unit and the ignition coil to test bench for further troubleshooting.

Fig. 6-20. Neon test lamp useful for checking CD ignition-coil primary voltage.

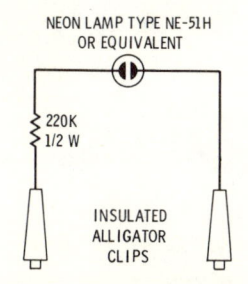

Locating a defect in a CD-ignition system on the bench requires a 12-volt power supply simulating the vehicle supply source, a temporary hookup to the ignition coil, and a switch for manually triggering the unit. Inverter operation can be checked by means of an oscilloscope, if desired. Voltage across the capacitor can be determined with a high-resistance voltmeter. A spark plug may be jumper-connected between the coil tower and the (−) primary terminal of the coil in order to visually check the spark output of the system. Once the system is connected and operating, opening the point-simulating switch should result in a spark. If a spark is not obtained, ordinary voltage and resistance checks should be made to localize the defective components.

MAGNETO-TYPE SYSTEM

Problems of a mechanical nature are more likely than purely electrical failures in these ignition systems. Prolonged exposure to moisture, wet grass clippings in lawn mowers, and ice and snow in snow throwers may result in corrosion of the components and wiring. Also, the harsh vibration and heat typical of small engines may loosen components, alter clearances, crack printed-circuit boards, or break wiring connections.

Careful disassembly, cleanup, and inspection will reveal most mechanical faults. Clearances are extremely important, especially between the laminations of the charge, trigger, and spark coils, and the permanent-magnet rotor. Check the manufacturer's specifications for these tolerances. Electrical problems will usually yield to continuity and resistance checks against the values provided in the manufacturer's literature.

Index

A

Advance mechanism, 13-14
Arc discharge, 10-11

B

Base, 24
Breaker points
  checking, 112-113
  curves, 16
  pitting, 113-114

C

Capacitive-discharge ignition systems, 69-102
  inverter-type, 71-99
    Crager "Power Pack", 81-83
    Delco-Remy, 83-92
    Delta Mark Ten "B", 76-79
    Jermyn Mk 2, 95-97
    Sydmur, 95, 98-99
    Thermo King "Thermotronic", 92-95
    Tri Star "Tiger 500", 79-81
  magneto-type, 99-102
  operation, 70-71
  troubleshooting, 125-132
Capacitor, energy stored in, 70
CD ignition systems; *See* capacitive-discharge ignition systems

Checking
  breaker-points, 112-113
  electronic ignition unit, 112-113
  ignition-coil primary voltage, 111-112
Chrysler electronic ignition
  troubleshooting, 119-122
Chrysler magnetically triggered ignition system, 50-58
  schematic, 57
Coil-to-distributor high-tension lead, 107
Collector, 24
Compression stroke, 8-9
Contact-isolator switching system, 42-44
  Judson, 42-44
Contact-triggered
  ignition circuits
    dual-transistor, 39-42
    complementary, 41-42
    simplified, 41
  ignition systems, 32-37
    Delco-Remy, 38-39
    Motorola, 39-41
  inductive-discharge ignition systems
    troubleshooting, 111-118
Conventional ignition, 13-19, 35-36
  graph, 36
  problems of, 14-19
Cragar "Power Pack" CD ignition system, 81-83
  schematic, 82

133

D

Dc-powered timing light, 117
Delco-Remy
  CD ignition systems, 83-92
    magnetically-triggered, 87-92
    schematic, 85, 89
  contact-triggered ignition system, 38-39
  magnetically triggered ignition system, 48-50
    schematic, 49
  unitized ignition system, troubleshooting, 125, 127-128
Delta Mark Ten "B" CD ignition system, 76-79
  schematic, 78
Diode, 22-23
Distributor
  cap, 106-107
  magnetic pulse, 86-88
  rotor, 107
Dual-transistor contact-triggered ignition circuits, 39-42
  complementary, 41-42
  simplified, 41
Dwell
  adjustment, 113-116
  computation, 61
Dwell-extender systems, 44-47
  SCR, 44-46
  Solitron, 46-47
  troubleshooting, 118-119
    electronic magneto, 119
    SCR, 118-119
    Solitron, 119
Dwell-tachometer, 113-114

E

Electrical-to-thermal energy conversion, 9-10
Electronic
  ignition unit, checking, 112-113
  magneto dwell extender, troubleshooting, 119
Emitter, 24
Energy conversion
  electrical-to-thermal, 9-10
Energy stored in a capacitor, 70
Engine, Otto-cycle, 8
Engine cranks but will not start, 108-109
Engine runs but misses at idle, 109-110
Engine runs fair with occasional missing, 110
Exhaust stroke, 8-9

F

Flame front, 11
Ford
  electronic ignition troubleshooting, 122-125
  magnetically triggered ignition system, 58-63
    schematic, 62
Forward bias, 22
Fuel-air charge, 8

G

General Motors/Delco-Remy unitized ignition system, 63-69
Germanium diode, 22

H

Heat sink, 23
High-tension lead, coil-to-distributor, 107
Holding current, 26

I

Ideal switch, 21
Ignition, 7-9
  circuits
    dual-transistor contact-triggered, 39-42
      complementary, 41-42
      simplified, 41
    single-transistor contact-triggered, 37-39
  coil, 13-16, 35, 60, 107-108
    primary voltage, checking, 111-112
    primary voltage waveform, 74
  conventional, 13-19
    problems of, 14-19
  Kettering, 13-14
  primary circuit, troubleshooting, 110-132
  secondary circuit, troubleshooting, 108-110
  spark, 8-9

134

Ignition—cont
  system(s)
    capacitive-discharge, 69-102
      inverter-type, 71-99
      magneto-type, 99-102
      operation, 70-71
    conventional, 35-36
      graph, 36
    inductive-discharge, 31-68
      operating principles, 34-39
      problems common to most, 103-108
      solid-state, troubleshooting, 103-132
    transistorized, 35-36
Inductive-discharge ignition systems, 32-68
  operating principles, 34-39
Intake stroke, 8-9
Inverter-type CD ignition systems, 71-99
  performance characteristics, 75-76
Ionization, 9-11

J

Jermyn Mk 2 CD ignition system, 95-97
  schematic, 96-97
Judson
  contact-isolator switching system, 42-44
  electronic magneto, 42-44

K

Kettering ignition, 13-14

M

Magnetic
  -pickup distributor, 48, 52, 59
  -pickup unit, 51, 64-65
  pulse distributor, 86-88
Magnetically triggered
  CD ignition system, 87-95
    Delco-Remy, 87-92
    Thermo King "Thermotronic", 92-95
  inductive-discharge ignition systems, 47-68
    Chrysler, 50-58
    Delco-Remy, 48-50
    Ford, 58-63

Magnetically triggered—cont
  inductive-discharge ignition systems
    General Motors/Delco Remy
      unitized, 63-69
      troubleshooting, 119-125
Magneto-type
  CD ignition systems, 99-102
    operation, 101-102
  conventional ignition system, 100-101
    troubleshooting, 132
Magwire, 76
Motorola contact-triggered igintion circuit, 39-41

N

Neon test lamp, 131

O

Otto-cycle engine, 8

P

Pickup coil, 49
Pn-junction device, 22
Points, breaker, 14-19
Pole-piece assembly, 49
Power stroke, 8-9
Programmable unijunction transistor, 28-29
P-type material, 21
PUT, 28-29

R

Radio-interference suppression spark-plug cables, 105-106
Regenerative switch, 26
Reluctor, 51-52, 64-65

S

SCR, 26-27
  dwell extender, 44-46
  troubleshooting, 118-119
Semiconductor, 22-23
  junction, 21-22
  theory, 21-29
Silicon
  controlled rectifier, 26-27
  diode, 22

135

Single-transistor contact-triggered ignition circuits, 37-39
Solitron transistorized dwell extender, 46-47
  troubleshooting, 119
Spark, 9-11
  plug(s), 9-11, 104-105
    cables, 105-106
      radio-interference-suppression, 105-106
    gap, 104
    time, 61
    voltage, 11
    waveform, 10
Stroke
  compression, 8-9
  exhaust, 8-9
  intake, 8-9
  power, 8-9
Sydmur CD ignition systems, 95, 98-99
  schematic, 98, 99
System runs but performance is poor, 113-116

T

Television resistance suppression wiring, 39
Thermo King "Thermotronic" CD ignition system, 92-95
  schematic, 94
Timer core, 49, 64-65
Timing, 116-118
  light, 116-118
    dc-powered, 117
  marks, 116-117
Transistor structure, 24
Transistorized ignition systems, 35-36
  graph, 36

Tri Star "Tiger 500" CD ignition system, 79-81
  schematic, 80
Troubleshooting
  capacitive-discharge ignition system, 125-132
  Chrysler electronic ignition, 119-122
  contact-triggered inductive-discharge ignition systems, 111-118
  Delco unitized ignition system, 125, 127-128
  electronic magneto, 119
    SCR, 118-119
    Solitron, 119
  Ford electronic ignition, 122-125
  ignition secondary circuit, 108-110
  magnetically triggered inductive-discharge systems, 119-125
  magneto-type system, 132
  solid-state ignition systems, 103-132
  the ignition primary circuit, 110-132
TVRS wiring, 39

V

Voltage
  -reference device, 23
  -regulator device, 23

W

Waveform, spark voltage, 10

Z

Zener
  diode, 23
  voltage, 23